Concrete Pipes and Pipelines

Unreinforced and Reinforced

Concrete Pipes and Pipelines

Unreinforced and Reinforced

N.G. Joshi

Alpha Science International Ltd.
Oxford, U.K.

Concrete Pipes and Pipelines
344 pgs. | 146 figs. | 59 tbls. | 51 photographs

N.G. Joshi (*Retd.*)
Chief Engineer
The Indian Hume Pipe Co. Ltd.,
Mumbai

Copyright © 2015

ALPHA SCIENCE INTERNATIONAL LTD.

7200 The Quorum, Oxford Business Park North
Garsington Road, Oxford OX4 2JZ, U.K.

www.alphasci.com

ISBN 978-1-84265-897-0

DEDICATED WITH LOVE AND AFFECTION TO
Late Shri Bahubali Gulabchand
Chairman and Managing Director
Shri Arvind R. Doshi
Joint Managing Director
Late Shri V. P. Limaye
Chief Engineer
The management, Technicians and Workers of
THE INDIAN HUME PIPE CO. LTD.
By N. G. Joshi

Preface

The pipelines are being constructed in ever increasing diameters, lengths and working pressures. Accurate and rationalized design basis are essential to achieve economical and safe design. Engineers have for many years resorted to semi-imperial design formulae. Much work has recently been done in an effort to rationalize the design of pipeline.

The book presents, the information collected during visits to various plants in advanced countries like France, Germany, Japan, etc., and discussions with the experts in the field. Although retaining conventional approaches in many instances, the aim of the book is to bring the most modern design and manufacturing techniques to Civil, Hydraulic and Production Engineers, who prepare Water Supply Projects and who use the pipeline and maintain it. Because of the sound theoretical background, the book will also be useful to Undergraduate and Postgraduate students. Many of the subjects such as standard bedding, Jacking pipe installation techniques are covered in the book.

It traces the history of mass production of concrete pipes. Gives various methods used for making concrete pipes and their merits and demerits in details.

Spinning technique has been discussed in more details, so that it will be useful to manufactures of RCC pipes and Piles also. Manufacturing problems and solutions are covered for the benefit of production personnel. Laying, bedding and design of rubber rings is included; rarely found elsewhere.

In short, it is the bible, for everyone, who is connected with Concrete Pipes for Water Supply, Drainage and roads. Information about the biggest and most modern plant at Baghdad will considerably help to modernize manufactures plants all increase the productivity, comparable to Japanese plants.

Numerous illustrative examples are given everywhere; hence it is expected that it will be useful to practicing Engineers in Water Supply Projects.

Pure water, in clean and healthy surrounding (free from water logged area) are the basic needs of human community. Pipelines fulfill both these requirements by supplying potable drinking water and collecting polluted and storm water. Pipelines made of different materials are in use for many years. But concrete pipe as a pipe material has not been replaced for drainage purpose by any other material even today. It has rugged construction and natural corrosion resistance. Gives solutions, to wide range of structural and environmental problems, Use of concrete pipes for conveying other liquids has dramitally increased in recent years.

Special chapter on the important aspects of the most modern plant, is included, together with photographs showing the important aspects of modern process for mass production.

Durability aspect of concrete pipe, together with the case study of 50 years old pipe, clears all doubts about the life of concrete pipe. It is easily more than 75 years.

N.G. Joshi

Acknowledgements

During my working period of 30 to 40 years, I had an opportunity to visit, over 90 factories in 25 countries, spread over the globe. The technicians in those factories have considerably helped me, to know the latest techniques in the manufacturing of concrete pipes. It is difficult to mention all names; hence I thank all of them collectively. However, I cannot escape, without mentioning the name of Chief Engineer Mr. Hiroshi Satoh of Nippon Hume, from Japan, with whom I also had an opportunity to work within the operation of Baghdad Plant.

I am thankful to late Shri V.P. Limaye, who was always a guiding hand behind me, for which I am grateful. My special thanks are due to my colleagues in R&D Division, of The Indian Hume Pipe Co. Ltd., Shri A.C. Patil, Shri K.H. Barde and Shri P. Ramachandran who considerably helped me in compilation and editing this book, without whose help, the book could not have come out.

Extensive knowledge I collected, is because of cooperation given by every member of The Indian Hume Pipe Co. Ltd. and the encouragement given by the management which is reflected in the book.

N.G. Joshi

Contents

CHAPTER 1

Overall Background of Concrete Pipe

1.1 INTRODUCTION

Pipes are vital, to the health and functioning of our community. A pipeline which is out of service even for a short period of time, causes extreme difficulty in conveyance and can have expensive consequences. Repairs and service difficulty can be minimised by directing more attention to the durability of the materials considered.

The durability of the pipe is as important as its ability to perform its structural and hydraulic functions. However prediction design, for durability cannot be made with the same degree of precision, as for structural and hydraulic performance. Although much progress has been made, in relating durability potential to such concrete properties, as pore structure, permeability and diffusion coefficient. Durability design still relies heavily on past experience. This experience consists of proven performance of over 150 years.

Shape of pipe is circular, because circular shape, has least perimeter, for conveyance of liquids. Apart from this, it is ideal from hydraulic point of view, as losses at corners such as a square, rectangular sections are completely avoided. Sometimes other shapes such as egg etc., are used where cross sectional area at the bottom is less as compared to top section. Such sections are very convenient when there is considerable variation is the discharge. When the discharge is less; the bottom section which has less cross sectional area, maintains the velocity. If the section is circular; at less discharge, the velocity will be considerably reduced.

1.2 HISTORICAL BACKGROUND

Over 2500 years ago, a civilization completed construction of one of the most renowned concrete water transportation system known to man. It conveyed water to densely populated regions. It was used as a supply main, for drinking water and for the first time as a tool to dispose of raw sewage. Today portions of this system are still functional. It was Rome's Cloacae Maxima.

Indian Archeologists claim that 2000 year before B.C.; there were pipes in India, made of stone, for conveying sewage in cities which had developed at that time. The knowledge and technology unfortunately did not continue further.

1.2.1 The Need

The need for sewers, water supply and drainage systems, has become more apparent during the course of human history. Practical methods have been developed, defined and refined over that span of time. Materials utilized have progressed from relatively simple applications of natural resources to durable and diverse precast concrete.

In England, France, Germany, Canada and the United States, sewage disposal methods advanced little beyond that of ancient Rome, until the 1840s. Then, in the early part of that decade, a system that connected all houses to sanitary sewer pipe maintained separately from storm sewers was constructed in Hamburg, Germany. The modern sewer system thus began to evolve.

In America, the first recorded concrete pipe sanitary sewer was built in 1842, in Mohawk, NY. The pipe was of bell and spigot design, with each section 28 inches in length. The inside diameter was 6 in (152.4 mm) and wall thickness was approximately 1 in. The pipe was installed by General Francis Spinner, to convey domestic sewage from his home to the Erie Canal. Spinner led a life of dedicated and innovative public service, as a Herkimer Country sheriff, a general in the state militia, a U.S. Congressman, and treasurer of the United States during the administrations of Presidents Lincoln, Johnson and Grant.

The pipe was excavated in 1982, for the 75th anniversary of the America Concrete Pipe Association and found to be in excellent condition, after 140 years of service. Many of the concrete pipelines installed in New England during the latter half of the 19th century are in use today. In Chelsea, MA, another early concrete pipe sanitary sewer was installed in 1869 at St. Louis, continuous to function satisfactory In 1868, a concrete pipe auxiliary sewer was installed in St. Louis, MO. In 1962, an examination showed the line to be in excellent condition, and it remained in service. A concrete pipeline installed as a combined sewer in St. Paul, MN, in 1875, is serving satisfactorily more than a century later. Between 1875 and 1888, the city installed more than 94,000 ft. (28560 M) of concrete pipes for combined sewers, which have provided 100 or more years of service. The growth of the precast concrete pipe industry in America over the past 150 years has followed the advance of technology. The increased need and concern for waste treatment, water supply, irrigation, drainage for railroads, highways and airports, has resulted in the construction of more larger pipelines.

The demand for sanitary and storm sewers, continued into the early decades of the 20th century. By 1915, most major cities had relatively extensive sanitary sewer systems. Many large cities, as well as smaller ones, used concrete pipe for sewer systems in U.S.A. In India pipes laid in 1926 are still in good condition.

1.3 SOCIAL BACKGROUND

1.3.1 From Pipelines to Roads

Early 19th century roads were improved only to the extent of having stumps and boulders removed. Many were impassable for wheeled vehicles in winter or during spring thaws. Travelers crossed small streams by fording and larger ones by ferry.

After 1820, the ideas of the Scotsman John L. Mcadam revolutionized American road building. The first American road built according to Mcadam's principle was the Boons Borough to Hagerstown turnpike in Maryland, completed in 1822.

A changing era, in transportation began with the shift from the need for local roads and streets to a more general, comprehensive, connected system. In 1893, the Office of Road Inquiry was formed in the U.S. Department of Agriculture to deal with local road problems. The increase in the number of automobiles between 1905 and 1918, from 50,000 to 6.2 million, indicated the direction of transportation development. Following 10 years of debate, the Road Act of 1916 established the concept of a national system of highways. The U.S. Office of Public Roads, formed in 1916, became the U.S. Bureau of Public Roads in 1991, but remained in the Department of Agriculture until 1939.

1.3.2 Growth

Local governments also were organizing highway departments. The growth of roads and highways resulted in a rapid increase in the use of concrete pipe. While the needs of the 19th century centered on public health and agriculture, the emphasis of the 20th century was to be on transportation needs. The demand for surfaced roads, reasonable grades and drainage, became essential considerations, in highway design. By 1930, all states were using concrete pipe in highway construction.

The use of concrete pipe for sanitary and storm sewers and for culverts, under highways and rail road's, grew steadily during the years before 1930. During the 1930s wide acceptance developed and more than two million tons of concrete pipes were produced. New developments in highways, during the

1930s, resulted in increased use of concrete pipe. Extensive use was made of concrete pipe on the 775 miles of the Pan American highway from Laredo. TX, to Mexico City, Mexico. Also constructed was the first of the modern freeways, the Pennsylvania Turnpike, which used more than 55 miles of concrete pipes.

1.3.3 Spectacular Growth

Between 1925 and 1930 the production of concrete pipe doubled from one to two million tons. The period from 1945 to 1965 was dominated by the needs of the automobile; the main thrust was the construction of the interstate Highway System. In 1956, the initiation of the $4 billion per year investment, produced an irreversible change both in transportation and in daily life. The development and construction of major airport facilities began during the 1930s. Between 1930 and 1940, almost 400 miles of concrete pipes were used in army, navy and municipal airports.

After the depression years and World War II, annual production doubled to four million tons by 1950. During each of the two succeeding decades, production increased by approximately three million tons, reaching a production level of more than 10 million tons annually by 1970. By the middle of the 1970s, the annual production approached 14 million tons with a market value exceeding $1 billion, representing approximately 1000 miles per month.

1.3.4 Still Busy

Even in 1980s, the concrete pipe industry in the United States is enjoying a period of prosperity. In almost every region, concrete pipe manufacturers are busier than they have been, since 1979. The exceptions to this are the areas that depend on farm income to stimulate the market, principally in the Midwestern region, and the areas that are dependent on the oil industry. In general, the economy that affects the concrete pipe industry is very strong. Housing starts have been up consistently, although the increases in recent months have been less dramatic than earlier. Non-building construction has remained strong, with increases in the order of two to three percent each month, for the economic community that the new tax structure will have a dampening effect on the construction industry for the second half of 1987.

1.4 TECHNICAL BACKGROUND

1.4.1 Present Condition

The Concrete Pipe Industry in North America to Europe has seen numerous developments in manufacturing process, automation and use of robotics to some extent. The developments in the area of hardware in Europe have been ahead of North America in some cases. Some Nordic countries and West Germany,

Switzerland have also introduced admixture such as silica fume, in Concrete Pipe Construction for better strength and sulphide corrosion resistance. The use of admixtures in North America is rather minimal compared to European practice and even in instances where admixtures are used in North America, there is usually fly-ash, a cement replacement.

In India, the market is huge but production of pipe is much less due to paucity of funds. In America and Europe, the cement used for pipe was around 2.0% of total cement produced in the countries. It is not more than 0.5% in India.

As on today, the developments in European countries are practically over, but still some developments are going on in America.

1.5 MATERIALS USED FOR MAKING PIPES IN THE PAST

Materials technology has evolved the most, among all aspects of pipeline engineering in the last 30 years. Many new materials and products appeared on the market, and an equal number of products and trade names disappeared. Some of the early and latest pipe materials are:

(a) **Stone:**   Pipes were carved through stone and used for drainage purpose. Few of such pipelines are in existence in India.

(b) **Lime Concrete:**   Well known oldest pipeline in Rome is still in existence.

(c) **Clay:**   The clay pipe industry is the oldest in perfecting the manufacturing technology. It is used mainly for drainage.

(d) **Cast Iron:**   This more than 150 years old. The pipes were used mostly for gravity main but the corrosion limited its use in present time.

(e) **Concrete Pipe:**   The first recorded concrete pipe sanitary sewer was built in 1842 in Mohawk in NY. It is discussed in length in later chapters.

(f) **Steel:**   To overcome the limited tensile stress, steel pipe was developed for pressure pipes. Invention of welding made it more popular now.

(g) **Plastic:**   The plastic pipe industry brought numerous developments in recent years. Resins such as Polyvenial chloride, Polybutadyne, Polyethylene, high density polyethylene and acrylontril butadiene system, entered the industry in smooth wall configurations.

(h) **Ductile Iron:**   Has better tensile strength and rigidity hence it has become very popular.

(i) **Composites:**   They have gained considerable momentum over the years using, primarily glass fibers either in the filament winding process or centrifugally cast process.

1.6 DEVELOPMENT OF CONCRETE PIPE

The 19th century brought a period of political consolidations and industrial expansion. Three areas of expansion during this period, produced the beginnings of the concrete pipe industry; public health requirements for water and sewage treatment, transportation and the agricultural needs for irrigation and drainage.

1.6.1 Public Health

Sewage disposal methods did not improve until the early 1840's when the first modern sewer was built in Hamburg, Germany. The cholera epidemics that savaged England around 1854, demonstrated the need for improvements in sewage disposal.

Many of the early sewers in America were built in small towns, financed with local funds.

1.6.2 Transport

One of the earliest rail road culvert was constructed near Salem, Illinois in 1854; an installation that was in service, over a century later.

1.6.3 Agriculture

Earlier drainage of farm and irrigated lands, consisted mainly of small open ditches which served to carry excess water away from low-lying areas, Concrete tile drain was developed in Holland, in the 1830's and introduced in America in the 1840's.

Scientific knowledge was first applied to the problem of irrigation and drainage in later half of the 19th century. As demand outstripped the production capabilities and the economics of mass production became apparent; new machines were developed to manufacture concrete pipes.

1.7 TECHNOLOGY AND MARKETS

The growth of concrete pipe industry was greatly influenced by related technical and market developments. Modern design and construction of sewers and culverts and the design and production of concrete pipe, have evolved from the basic work of the last 100 years. This activity included:

- Development of hydraulic and hydrological theories
- Concepts of loads on pipe
- Standards for materials and tests

1.7.1 Hydraulic and Hydrological Theories

The basic theory of modern pipeline design was developed over the latter half of the 19th century. Of principal interest were the studies, to determine head loss from pipe wall roughness. These studies formed the basis for determining pipe size. The results of these early studies, beginning in the late 19th century, are still being applied. Most important among them are:

- Darcy and Weisbach in 1857, extended Chezy's open channel flow formula to pipe flow.
- Hazen and William in 1902, developed an equation for flow in pressure and gravity pipes.

1.7.2 Loads on Pipe

During the first three decades of the 20th century, researches in Iowa State University developed and tested, a theory for estimating loads on buried pipes. The original concept was advanced by Marston-Talbot and the theory was developed by Marston and Anderson and published in 1913. A.Marston was joined by M.G. Spangler and W.J. Schlick and continued the work on evaluation of design loads. In 1930, Marston published "The Theory of External loads on closed conduits in light of the latest Experiments" which presents the theory in the present form. During the same period, the three edge bearing test was developed as a method of evaluating the strength of rigid pipe. Other IOWA reports include Schlick's test on pipe on concrete cradles and Spangler's classic report on supporting strength of rigid pipe culverts, which still serve as the principle design theory.

1.8 STANDARDS

The quality of Concrete and Concrete Pipes received extensive attention throughout the early years of the 20th century. The major forum for these studies was the American Society for testing of materials ASTM. The history of Concrete Pipe Standards began virtually with the founding of ASTM in 1898. By 1904, eight technical committees were formed; of these committee C-4 was the fore number of committee C-13 on Concrete Pipe.

In 1918, the ASTM Joint Culvert Pipe Committee was formed and consisted of representatives of the American Association of State Highway Officials, American Concrete Institute, American Railway Engineering etc. The purpose of this committee was to develop a standard for reinforced concrete culvert pipes.

Organizers of the association, considered the development of standard and uniform quality to be of greatest importance to new industry. Most of early discussions at Annual Meetings were concerned with ways and means of improving quality in concrete drain tile and by 1914 a specification for drain tile was developed.

Perhaps the most significant product change during the period came with the introduction of reinforced concrete pipe. Reinforced cement was first used in France in 1896 and the concept was brought to America in 1905. Most famous specification C-76, for reinforced concrete culvert storm water, sewer pipes was formed in 1930.

Internationally recognized Specifications on Concrete Pipes came out in the following years.

Country	First published	
• U.S.A	ASTM C-76	1930
• British	B.S.-556	1934
• Australian	A.S-A 35	1937
• Indian	I.S 458	1956
• Japanese	JISA 5303	1972
• European	ENC 640	1994

1.9 ADVANCEMENT OF TECHNOLOGY

While much of the theory had been developed prior to 1930, subsequent research and standardization contributed greatly to reinforcement of earlier standards.

During the last 70-80 years a variety of pipe manufacturing process have been developed, with a resultant significant increase in pipe quality, longer length and ever increasing pipe diameters. The spun and rolled processes are Australian inventions and have been widely licensed overseas, a recognition of Australian quality.

Even before the era of Hume of Australia, reinforced concrete was used for conduits under comparatively small pressures. But these were so massive and costly that in industrialized countries like Europe and USA, practically all pressure pipes were of Ferrous metal. Australia prior to producing iron and steel, had no choice but to import such pipes at relatively great expense. The precast concrete pipes were then very crude. Virtually could not reliably resist pressure heads exceeding 15-20 meters.

Between 1910 and 1920 Hume Brothers in Australia developed spinning process. Reinforced, factory made pressure pipes, soon became commercially viable. Total of Concrete Pressure pipes made in Australia up to 36″ (900 mm) dia. and tested to 235 feet's exceeded 2 million feet by 1930.

Rubber ring joint was introduced in 1934 in place of caulked joint. This was a, major improvement in many respects as it greatly enhanced the advantages of reinforced concrete pipes for widening range of pressures. Initially the, pressure pipes were jointed with steel collars. Those were mortar caulked with medium of longitudinal freedom provided by lead-caulked, joint about every 4th joint. This was marginally satisfactory for buried pipes but in portion exposed to severe frost, hot sun and drying winds, even if all the joints were of lead, they would allow only small faction of expansion and contraction as afforded by rubber ring joints. The resulting longitudinal forces caused transverse cracks. From the leakage and subsequent autogenous healing occurred so often, that strips of calcium carbonate (marble) created a Zebra-like appearance.

The reinforced concrete pipe appeared in Patents in Germany in 1879 and was used in Sewerage Construction in Paris in 1892. First recorded use, in USA was in 1906, in Australia in 1910, and in Indian around 1925.

1.10 THE FUTURE

Modern man cannot live without adequate sanitary sewers, storm sewers, and culverts. Concrete pipe is strong durable, economical commodity that has made the modern infrastructural system possible and will continue to be installed and function for future civilization. Though the demand in developed countries is now reduced; in India and other under developed countries. It is still huge.

1. Section of concrete pipe from an underground aqueduct built by the Romans around 80 A.D. between Fiffel and Cologne. Reproduced from Concrete Pipe Handbook[*] (American Concrete Pipe Association, 1980).

Photograph 1

[*]John Duffy and Mike Bealey – ASTM standardisation news September 1987.

WALTER HUMES
*"I was searching for some proposition where the
centrifugal action might be turned to good account..."*

Photograph 2

Design Team

CHAPTER 2

Different Methods of Mass Production of Concrete Pipes

2.1 EARLY PIPES

The first pipe is, said to have been made by Mr. Monier, a French man, by tamping concrete in the annular space between inner and outer mould. The pipe was of one meter length and used as flower bed. Even today such pipes are made in villages for irrigation.

After the public became conscious of the need for sanitation, concrete sewer pipe was developed during 19th century. Many installations of concrete pipe had been made prior to 1880 and their durability characteristics soon become apparent.

The large sewers constructed in Paris during the middle of 19th century were built of rough stone heavily plastered with cement mortar on the interior.

Many of concrete pipelines installed in New England during latter half of 19th century are still in use today.

In 1988, a concrete pipe sanitary sewer was installed in St. Louis, Missouri. An examination of this in 1968, showed the line to be in excellent conditions and it remained in service.

All these pipes made even by crude methods, indicated the durability of concrete pipe. These were basically made by a stationary mold, placing concrete in the annular space and tamped it.

2.2 DEVELOPMENTS IN MANUFACTURER

20th century brought new technology, new standards and new rates of growth for concrete pipe industry. The need to improve quality, and production capabilities was recognized. By 1914, a specification for drain tiles was developed. Calendar of developments of modern machines for production of concrete pipes is as given on next page

- Machines for making Concrete Pipes were developed both by
  Mc Cracken and Zeidler in Iowa using packer head system   1905
- Quinn wire and Iron work of Bonna, Iowa made Machine
  using steel tamper sticks, Both inner and outer forms revolving   1906
- Humes, Australia developed the process of Centrifugal Spinning   1910
- Roller suspension in Australia.   1947
- CEN – VI RO, U.S.A   1950

The main differences in technical points in these processes are, the position of casting and method of compaction, some are casted in vertical position and other in horizontal position.

2.2.1 Tamper Process

It uses direct mechanical compaction to consolidate the concrete mix. Inner or outer forms are placed on rotating table and concrete mix fed into the annular place as the form is rotating the tampers are raised automatically. There are usually multiple tampers so that the mix on each side of reinforcement can be compacted. The pipe is removed from the machine with either the inner or outer forms are moved to cured area where the remaining form is removed Fig. 2.1

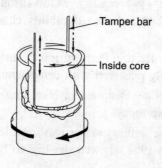

Fig. 2.1   Tamping

2.2.2 Packer Head Process

As the name suggests the consolidation of the concrete is by a packer head. Mc Cracken machine which works on this principal is shown in Fig. 2.2

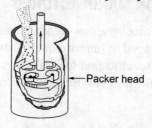

Fig. 2.2   Packer head

The machine is set up with desired size attachment. An assembled form is placed on the off bearing position on the turn table. After rotating through 90°, the turn table automatically stops and indexes the assembled mould to the pipe making position.

The machine hopper is lowered on the top of the mould while the roller head and packer shaft continue downward. Hydraulic pressure is aromatically applied to the hopper to prevent distortion of tongue formed during production sequence.

The downstroke is controlled both electrically and hydraulically, to provide two different speeds for the down stroke of the packer shaft. As the roller head approaches the curing pallet located in the bottom of the assembled mould, automatically slowdown and continues downward, then automatically stops at the correct lowered position.

The conveyor with concrete is then started and after adding a sufficient amount of concrete to form the socket of pipe, the conveyor is shut off. Additional water is then added to the socket. At this point the packer shaft rotation is started and the socket down unit is raised. As the socket down unit engages the curing pallet, the pallet begins to rotate and vibrate, the completion of cycle is indicated when the preset timer stops the rotation and vibration of the curing pallet.

The conveyor is started again and the packer shaft starts moving upward. As concrete is being fed to the mould, the roller head distributes, compacts and trowels the inside of the pipe – all in one operation (Fig. 2.2). Throughout the lift of roller head, the feed of concrete is maintained to avoid any variation in the packing of the pipe.

A second finishing and polishing pass is then made. The hydraulic pressure applied to the hopper is automatically released. The hopper lifts s off the form and continues to lift with the roller head. The lift automatically stops at the correct height to enable the turntable to rotate. Once again the turntable is rotated through 90° indexing the second assembled form to pipe making position and complete pipe, to off bearing position. The entire production cycle is then repeated.

The pipe with the form is then taken away from the machine and moved to the stripping area by a crane. The form is opened there and lifted from the pipe. The crane then lowers the form on a new set up and returns the assembled form to the pipe making machine.

The uncured pipe are allowed to obtain initial concrete set, then are cured overnight under drop curtains, using low pressure steam. All these operations are illustratated in Fig. 2.7.

2.2.3 Vibration

The pipe is cast in vertical position. Concrete with consistency of moist earth is fed into annular space between the moulds. The concrete is compacted by a vibration core which is called internal core vibration system. Beside this, the pipe is pressed tightly by a hydraulic press which at same time form the upper end of pipe,–the spigot end.

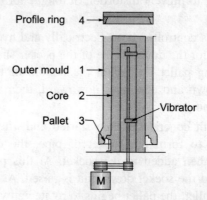

Profile ring 4 →

Outer mould 1 →

Core 2 →

Pallet 3 →

Vibrator

M

Fig. 2.3 VEHY machine

After casting, demoulding take place immediately and the pipe is supported in the spigot and socket ends during setting of the concrete.

The vibration originate from the vibrator or from the machine which is fixed in the center of the core. After filling, spigot ends in pressured at the same time as the profile ring in socket, forming the spigot end profile. Pressure in then stopped but the profile ring remain on concrete.

Vibration is stopped, then demolding commences. The mould is pulled off the core from above or below. The pipe now stands on the pallet. The pipe then remove off the core, from above. A set ring is place on the spigot end during the period of setting. Next day the pallet is removed.

2.2.4 Centrifugal Spinning (Horizontal Position)

In this process a steel mould is related in horizontal position on rotating wheels. Initially at low speed, concrete is deposited evenly inside the mould. The speed is increased; the centrifugal force generated due to this, presses, the concrete against the sides of the mould. It compresses the concrete by extracting water (having less specific gravity) and produce dense concrete. Initially the water cement ratio is high which gives workability to concrete and final water cement is as low as 0.28 to 0.3 hence it produces a dense durable concrete.

This process was inverted in 1910, by Water Hume of Australia and popularized throughout the world. Now, after development of low amplitude and high frequency vibrators, pipes are made in a vertical mould where the concrete is always in compressive mode. As the initial water cement is low the concrete achieves strength very rapidly and hence the pipe can be demoulded immediately, hence the productively is high as compared to spinning; also the same mould can be used again and again and hence the investment on mould is also less.

This process was inventory by water Hume of Australia. similar, Process with slight difference were also developed by Vianini of Italy, Bonna of france.

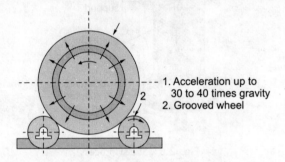

Fig. 2.4 Spinning under vibration

2.2.5 Roller Suspension

Another method is called the roller suspension process. In this process cylindrical moulds with cast steel end rings are hung and rotated on a heavy shaft. Concrete is fed into the rotating mould and packed against its inside surface by a highly efficient combination of centrifugal force, rolling compression and vibration. It is claimed that compression of very dry concrete (W/C ratio of about 0.28) by this method ensures strength, in excess of those achieved by other traditional core making methods.

Fig. 2.5 Roller suspension process

2.2.6 Centrifugal Spinning (Vertical Position Stussi Process)

It differs from all other centrifugal spinning machines, it operates with the mould vertical and uses comparatively dry mixes, a water cement ratio of about 0.35 being the aim. This is made possible by the use of a distributor table which moves up and down inside the pipe form, as regulated by the operator. This distributor table also rotates at a greater speed than the form, and mixed concrete is allowed to fall on to it from above. As soon as the concrete strikes the table it flung out against the sides of the mould with such force that it remains in position with particularly no subsequent movement. It is claimed that in this method there is no segregation and the cross section of the finished type show perfect distribution of the aggregates throughout the wall thickness.

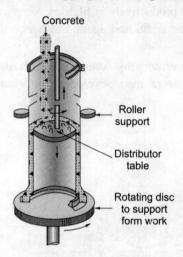

Fig. 2.6   Stussi process

2.2.7   Productivity

Productive of VEHY Process is given below

Table 2.1

Productivity of Process									
Pipe Diameters cm φ	30	40	60	80	100	120	140	160	180
Meter per hour. Based on pipe length of 2.5 m.	20	20	18	15	12	10	7	6	5

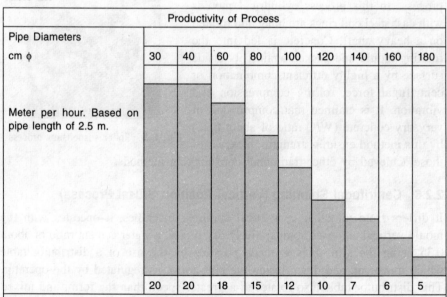

When shorter pipes are produced, a minor output may be excepted.
Reinforcement will mean a reduction of the figures.

2.2.8 Productivity of Various Processes

Table showing the comparison of productivity and labour force required at the same level of machanisation is given below. Basically, figure of output and labour are for 900 mm dia NP2 class socket and spigot pipe. Only the pipe making operation is considered but not those of reinforcement, curing etc. When shorter pipes are produced, a less output may be expected. Reinforcement will also mean a reduction in the figures.

Table 2.2

Process	Size	No of 2.5 M Long pipe in 8 hours shift	No of workers
Spinning	900	20	12
Roller Suspension	900	45	10
Vihy Transmatic	900	90	8
Mc cracken	900	144	10

The figures hardly need any comments.

2.3 COMPARISON BETWEEN THE SPINNING AND VIBRATION CASTING PROCESS

Table 2.3

Item	Spinning process	Vibration casting
1. Basic Difference:		
a. Compaction	By centrifugal force with extraction of water. And reducing the water cement ratio.	High frequency vibration combined with pressing.
b. Water Cement Ratio	Initial: less than 0.45 Final: i) around 0.35 to 0.30 depending upon the mix design, process, speed and time of spinning/ ii) difficult to judge the final w/c ratio and hence the strength.	Around 0.30 i) Final 0.30 same as initial ii) Easy to judge the strength as exact strength can be had from cubes. iii) Strength under perfect control.
c. Suitable Thickness and Diameter Range	Min. thickness of 25 mm. Max. thickness-Normally up to 120 mm. For thickness 70 mm and more spinning has to be done in two layers to achieve the required strength specially at socket.	Min. thickness is 50 mm. Max. Thickness up to 250 mm easily possible.
d. Possibility to Manufacture the Unreinforced Pipes	Possible, but achieving correct socket dimension is difficult.	Possible, diameters of socket are accurate.

2.	Process Time*	Min: 25 min. for one run for 150-300 mm dia pipes. -6 to 8 Nos. at a time. For Higher dia like 1200 mm time is around 1.25 Hrs. (75 minutes) Spinning can be stopped only after required strength is achieved.*	Less than 7 mins. for 300 and less mm dia pipes. Making multiple pipes together is advantages. e.g., 3 to 4 in that case 7 mm for 3 pipe line at a time. For 1200mm dia. Pipes time is 15 to 20 minuts. In process, the pipe is in vertical position; hence there are only comp-ressive stresses on the pipe.
3.	Demoulding	After 24 hours, if heat curing is not done.	Immediately after casting.
4.	Requirement of Moulds	Same number of moulds as number of pipes to be made, in a shift	Only one mould is requi-red for one size
5.	Joint	Mortar or Rubber Ring	Mortar or Rubber Ring
6.	Accuracy of Spigot and Socket for R.R. Joint	Spigot better, but difficult to achieve the required socket dimensions if mould is in one piece.	Very accurate. Ideal for R.R. Joint
7.	Environmental Pollution	The water coming out, is dirty and surroundings are difficult to maintain clean	Clean and healthy surroundings.
8.	Mix Design	During feeding and spinning, further mixing takes place. Hence Mix design accuracy not very predominant.	Has to be perfect. No adjustment during process is possible. Hence, mixer must be very good.
9.	Cage Maintaining Cover	Cage is likely to be disturbed during concrete feeding operation. Hence, cover is non uniform if spacer blocks are not used.	Not distributed as the feeding of concrete is in vertical position.
10.	Material Requirements	Slightly less cement is required.	Slightly more cement is required as fines are more in mix.
i.	Coarse Aggregates	Special grading, allowing water to come out during spinning is necessary, little mixing takes place during spinning.	Must be perfectly graded.
ii.	Fine Aggregates	Grading of sand must be good to achieve perfect spinning. Use of crushed sand not easily possible.	Crushed sand can be used. Mix grading is on finer side.
iii.	Cement	Little higher cement is required than normal concrete, as during spinning some cement comes out with water. 43 grade cement is preferable. Difficult to maintain the shape as it gets disturbed during feeding of concrete.	No loss of cement. For the required strength minimum 43 grade cement is preferable. As the mix is on finer side little more cement is required than spinning.

iv. Cage-Possibility of using elliptical or quadrant reinforcement	Not so easy to use elliptical or quadrant reinforcement. Proper spacers have to be used.	Easily possible to use elliptical or quadrant reinforcement.

* In horizontal spinning, the top portion of concrete is in tension, hence compaction has to be done till the required tensile strength is achieved to with stand tensile stresses.

* In spinning, time required for hardening depends upon the quantity of water to be taken out. If some admixture is used, the initial water cement ratio can be reduced without affecting Workability.

2.4 CHANGE OVER FROM SPINNING TO VERTICAL CASTING

Till 1960 most of the manufactures who were making pipes by spinning, changed over to vertical vibration, obviously due to high productivity of vibration process especially in Europe and U.S.A. Presently due very low demand, all these units are closed especially in Europe, because the developments there are practically over.

However certain developed countries, like Japan and Switzerland never gave away spinning process and still they use it. Recently with the introduction of chemical and mineral admixtures very high strength spun concrete is produced with resultant water cement ratio of as low as 0.2. The strength of concrete achieved is as high as 110 Mpa. A detail of mix used in that case is given below per cum.

Cement	450 Kg/m^3
Coarse Aggregate	1413 Kg/m^3
Fine Aggregate	410 Kg/m^3
Silica Fume	45 Kg/m^3
Water Cement Ratio	0.2 (90 Kg)
Super plasticizer	6.5 ltr/m^3
Concrete strength at 28 days	110 MPa

2.5 PROCESSES MOST COMMONLY USED NOW ARE

(a) Horizontal spinning.

(b) Vertical casting by vibrators.

2.6 INDIAN SCENARIO

Photograph 3 Seth Walchand Hirachand

In India, concrete pipe by spinning process was produced and popularised on large scale by Seth Walchand Hirachand around 1926, through The Indian Hume Pipe Co. Ltd.; with

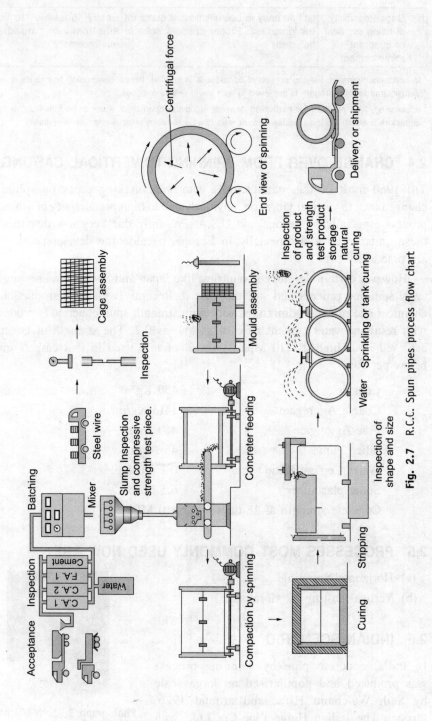

Fig. 2.7 R.C.C. Spun pipes process flow chart

the help of Humes of Australia. The pipe is now very popular throughout the country. Every year, thousand of tones of concrete pipe are made. Demand for the concrete pipes will increase as most of the states do not have adequate underground drainage facilities.

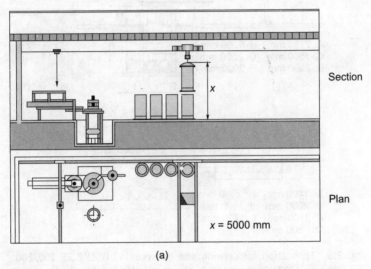

(a)

Fig. 2.8(a) The picture shows the removal of formwork and a crane. Two forms of outdoor use, it can be retracted into the machine with a small electric train, the second outer mold, while the crane transported the previously-made pipe and demolded

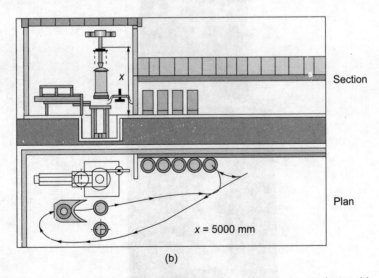

(b)

Fig. 2.8(b) The picture shows the demoulding with a crane next to the machine, while a tube ride the trolley to the finished pipe space. This method requires a much lower ceiling height at the parking lot

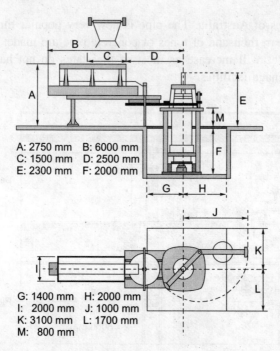

A: 2750 mm   B: 6000 mm
C: 1500 mm   D: 2500 mm
E: 2300 mm   F: 2000 mm

G: 1400 mm   H: 2000 mm
I: 2000 mm   J: 1000 mm
K: 3100 mm   L: 1700 mm
M: 800 mm

Fig. 2.9 Foundation for drawing and Massskize VIHYSIMPLES 100/200. When VIHYSIMPLEX 100/250, the foundation pit 50 cm deep

Fig. 2.10 Vertical casting process. Outer mould is lifted when concrete is stiffend

CHAPTER 3

Horizontal Spinning Processes
and P.V.C. Lining

3.1 WHAT IS CENTRIFUGALLY CAST CONCRETE AND ITS ATTRIBUTES?

It is a novel method of compaction of concrete. In technical terms, it is a compaction of concrete by water extraction. It has been used for a long time in making precast poles and pipes. In this process high workability (high W/C ratio) concrete can produce high strength concrete, which is not practicable without admixture. On normal concrete during spinning, the fresh concrete is subjected to very high centrifugal forces, that compact the material, against the interior of the steel mould and expel excess water in the mix. The overall porosity of the material is therefore reduced and as a result, the hardened concrete is exceptionally dense and have high strength.

The higher strength is attributed mainly due to lower W/C ratio and higher density of concrete after spinning. Due to its lower permeability, the material is highly durable and can withstand harsh environment. The dense concrete provides good protection for the steel reinforcement against ingress of chlorides and other harmful chemicals. Concrete covers as small as 16 mm have been used in many pipe structures across the world, giving excellent performance over many years of service.

The main advantages are that a high workability concrete can be placed relatively inexpensively, the final product has a considerably lower W/C ratio and can therefore be of high strength and good durability. Therefore, freshly processed product can be stripped of its mould and can be reused early. The principal examples of water extraction are the hydraulic pressing of kerbs, paving slabs and spinning of concrete pipes.

The essential requirements of the concrete pressure pipe are **integrity, impermeability** and **high tensile strength**. Spun concrete process is able to compact without a flaw, mixes with water cement ratios initially below 0.40.

After expulsion of water by spinning, the final W/C ratio may be around 0.25 to 0.3. Concrete strength up to 85 MPa can be easily achieved.

Its Attributes

Although very little information is available in the literature on the properties of spun concrete, there are several indications and unpublished experiences that allude to improved material properties. Abeles reported some of his experiences in the design and manufacture of centrifugally cast concrete masts and pipes in Europe, and pointed to the improvement gained from the spinning operation. No specific test results were given, but the fact that the spun concrete is much stronger than conventional (non spun) concrete is stressed throughout the paper. Several other studies discussed centrifugal casting, as it relates to the manufacture of pipes and alluded to the advantages of the method. The Prestressed Concrete Institute, The American Society of Civil Engineering and the American Society of Testing Materials, have also realised the importance of this growing industry and published report and standards addressing the design and use of concrete poles with special attention given to the spun poles.

It has been shown by U.S. Bureau of Reclamation that concretes made by spinning are:

(a) About thousand times as impermeable as good (engineering) quality structural concrete placed in situ.

(b) Pipe density of spun concrete is higher than conventional concrete.

(c) **Only concrete of this standard, have the direct tensile strength to resist, without cracking, the primary stresses in pressure pipes. Therefore cracking, inherent in normal reinforced flexural member is impermissible.** The stresses in the concrete pipe wall at test pressure, do not usually exceed 300 to 400 psi. (21 to 28 kg/cm^2) Since cracking is negligible, the stress in steel can be only that in concrete, multiplied by Es/Ec. At test pressure, this might be up to 3000 psi (211 kg/cm^2) or increasing to 15000 psi (1055 kg/cm^2) at concrete failure. This is a small fraction of ultimate tensile strength (about 80000 psi [562 kg/cm^2] to 100000 psi [703 kg/cm^2] of steel. Whatever the amount of reinforcement in the section, having to remain virtually uncracked, the concrete definitely provides 90 to 100% of the tensile resistance. **Thus the traditional concept of cracked concrete is quite inapplicable to pressure pipes.**

Since the steel is doing so little, why do we provide so much? One answer is that, in order to specify the amount, the clause reverts to the orthodox assumption that the concrete is incapable of resisting tension. This is illogical, since the concrete will then crack and releases the pressure which causes the tension. As a device, however, it has the merit of needing no new skill in design to ensure the superior performance under shock load for which R.C.C. pipes are noted.

(d) Their behavior differs radically from that of brittle pipes (Plain Concrete, Cast Iron, Asbestos Cement, Ductile Iron). These subjected to accidental pressure beyond their capacity, burst irrevocably. But an over stress in R.C.C. pipe can crack and leak until it has shed the excessive pressure.

(e) At the instance of cracking, the reinforcement is invoked. It holds the sides of the crack together tightly, or even surrounding moisture ensures autogenously healing. The marble which soon fills the crack is usually stronger than the original concrete. This action is ensured by suitably moderate stress in reinforcement so that it prevents opening of the crack and hence, any tendency to stretch the steel is of negligible dimension, of the crack width.

(f) Criteria for satisfactory shock (water hammer) resistance are well established. Hence the cross-sectional area of hard drawn wire spirals in a length 'L' of pipe with internal diameter 'd' for test pressure 'P' is

adopted as $\dfrac{PdL}{4000}$.

These findings, no doubt establish the supremacy and suitability of spun concrete for pipes. **However, as the properties of spun concrete are not known from the standpoint of strength nor permeability, very little information is available in the literatures. Improvements in the physical properties of the spun material do occur, but cannot be estimated reliably to incorporate in the design. The situation is further complicated by the unavailability of standard method for casting and testing spun concrete cube.**

3.2 HORIZONTAL SPINNING PROCESS AS APPLIED TO CONCRETE PIPES

Principle

In this process, concrete is compacted by huge centrifugal force 30 to 40 times the acceleration of gravity, in a horizontal rotating mould; with steel cage, to form the body of the pipe, Pipes of diameter 150 mm to 3000 mm, in 2.5 mtr lengths are generally made Fig. 3.1, following are the main features of this method of production.

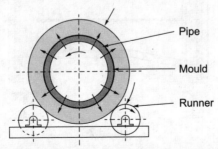

— Pipe

— Mould

— Runner

Fig. 3.1 Horizontal spinning machine

(a) Concrete body is compacted tightly. There are no bubbles and so absorption is small.

(b) Surplus water is removed by centrifugal force. Resultant water cement ratio of concrete will be small. The pipe body will therefore be strong and durable.

(c) The outside surface of pipe is more beautiful than any other method of production. The inside surface is very smooth, accordingly resistance to water flow is less.

(d) The manufacturing facilities are rather simple and trouble free. Manufacturing does not require high technical skill.

(e) Machinery is simple. Adjustment of machinery is not necessary for certain range of inside diameters; so occasional change over of inside diameter will not lower productivity.

The compressive strength of concrete manufactured in this way is more than 400 kg/cm². High class blending easily produces concrete with compressive strength of over 600 kg/cm². Various types of pipe strengths could be had by combining concrete strength and amount of steel reinforcement. Recently, by using special blending materials, like micro silica. It has become possible to greatly increase the strength of pipe against external load and internal pressure. Detail Speeds at which different diameter of pipe to be spun, is given in Enclosure B. All sorts of pipe joints can be easily made by selecting appropriate mould.

3.3 MAIN STAGES IN PRODUCTION OF CONCRETE PIPES

3.3.1 Feeding Concrete in the Mould and Compacting it

The machine basically consists of two runners on which the mould is supported with end rings on both sides. The runners are fitted on a strength which supplies power from a variable speed motor show in Fig. 3.2

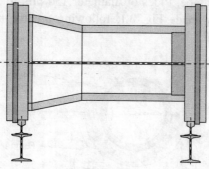

Fig. 3.2

Initially the mould in rotated at slow speed and concrete is fed into the mould uniformally throughout the length. The thickness of the concrete is controlled by the end ring. When pipes with socket are to be made, socket former fitted on one side. The concrete in fed either manually or by a belt conveyer. The initial speed is such that when the uncompacted concrete goes on the top, it should not fall down, usually the speed in 3 times the gravity.

When the feeding is completed and concrete is leveled, the speed is gradually increased, during this time the concrete which is in fluid form gets adjusted due to centrifugal force acting from inside face of pipe. Some excess water comes out.

The speed is then increased to the maximum spinning speed, which is usually 20 to 30 times. The gravity high speed is maintained for 5-10 minutes, depending upon the thickness of pipes. At this time the concrete is compacted tightly by removing air bubbles and excess water. The compaction is so effective that the outside surface of pipe is beautiful than any other method of making pipe.

The initial water cement (w/c) ratio is around 0.4 to 0.45 and the final or resultant w/c ratio is around 0.2 to 0.25, at this w/c ratio, the concrete is very strong.

At the end of full speed spinning, water which is coming out due to Squeezing is removed by a flat steel bar or by a water jet. The internal concrete surface is polished with steel flat or sleeker i.e., a belt piece fixed to small dia pipe, if required additional cement may be used to get good polish.

When the polishing is completed and complete water inside the pipe is taken out, the rotation is stopped. The concrete pipe is lifted with the mold either by an electric hoist, crane or chain blocks and with traverses and placed in curing area.

Generally pipes from dia 350 mm to 3000 mm in 2.5 meter length are made by this process. Pipe from dia 150 mm to 350 mm dia are made in 2 meter lengths. During spinning such 6 to 8 pipe are spun at a time and bigger diameter pipe in 2.5 m length, from 1 to 4 pipe at a time depending upon the weight.

The beauty of this process is that required member of pipe can be easily taken at a time. When the diameter changes the distance between the runners can be easily changed. Hence considerable time, in adjusting the machine when diameter changes can be saved.

The following table gives diameter ranges and the member of pipe that can be taken in one run, no of ranges and the member of pipes produced in on 8 hours shift are given on next page.

Table 3.1

Size of pipes	Range of dia	Quantity produced in one run	No of runs	Quantity produced per shift
Small diameter	150 - 350 mm	8 - 12 Pipes	8-10	80 - 120 Pipes
Medium dia.	400 - 1000 mm	3 - 5 Pipes	6-8	20 - 40 Pipes
Larger dia.	1100 - 2000 mm	2 - 3 Pipes	6-6	12 - 20 Pipes
Extra large	2000 - 3000 mm	1 - 2 Pipes	6-6	6 - 12 Pipes

3.3.1.1 Difficult thing to achieve in horizontal spinning with socket ended pipes for rubber ring

Joint for forming a socket in the pipe.

Main peculiarity of spinning process is that initially we can put more water and during spinning take out extra water, so that the resultant w/c is low and concrete is harder. However, at the socket portion of the pipe, we are unable to take out water coming out in the portion BC of the socket Fig. 3.3. Hence this water remaining.

To avoid this, concrete should be fed in two layers. In the first layer, concrete should be fed partially XYZ as shown in the Fig. 3.4. Then the spinning should be done, for first layer, water removed to the extend possible from socket. As the water from the first layer QR Fig. 3.4 as practically removed concrete in this portion hardens without much difficultly shown in Fig. 3.4. In second feeding the concrete should be fed slowly, so that no air remain in socket portion and it is filled with concrete, during second spinning, as the quantity of water coming out is very small, it does not matter much, some water does come out at the end of socket former as shown Fig. 3.5.

Main thing is that, the thickness of concrete in the first feeding, should be such that adequate space should be left over the concrete in socket. This space, should also be not more than that required for accommodate water coming out during second spinning, but this water is much less, hence it will be able to produce better socket. Hence it is preferable, to do feeding in two layers when the thickness of concrete wall is more than 70 mm.

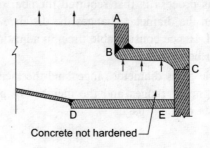

Concrete not hardened

Fig. 3.3

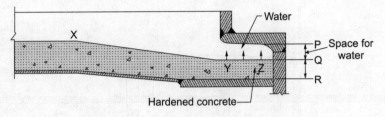

Fig. 3.4

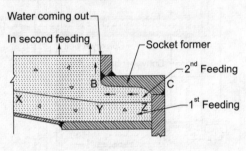

Fig. 3.5

3.3.1.2 *Concrete conveying system*

Concrete is conveyed from the mixer to the various pipe making machines by belt conveyer or a bucket which has a automatic opening – closing lid as shown on next page.

3.3.2 Demoulding and Assembling

After the concrete has attained the required strength, which is usually after 20 hours or 5.6 hours if steam curing is used. The mould is opened out and the pipe is taken out and carried to curing area for further curing.

When the pipe is removed the mould is cleaned and reassembled for use again.

Most important thing is that, this operation should be done quickly so that mould can be reused agent. However it, depends upon the design of mould and the process followed for demoulding and assembling of the mould. This is discussed in "Mould" under equipment.

3.3.3 Reinforcement Cage Making (It is Covered in Separate Chapter)

3.3.4 Curing

This is most important operation for any concrete product. Main thing is, pipe should not subjected to any blowing wind run etc., when the pipe is in the

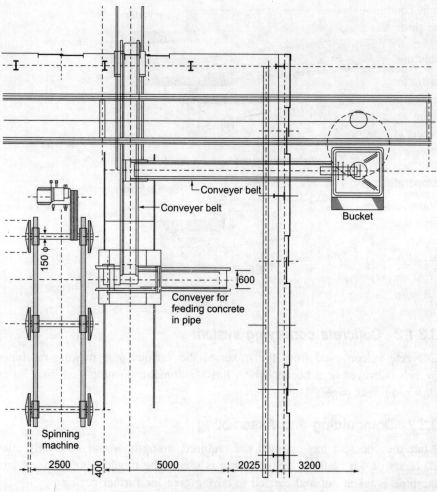

Fig. 3.6

mould wind should not be allowed to blow through the pipe, as it evaporate the water from the surface of pipe. Hence ends should be closed to prevent blowing of wind and the mould from run hence in should be covered.

After the pipe is taken out from the mould. The pipe should be in humid conditions. This can be created by having sprinklers or keeping the pipes in water tank. More details about curing are given below.

Importance of curing

Like other chemical reactions, hydration of cement depends on temperature and time but hydration also depends on presence of moisture.

As concrete hardens, it becomes stronger, resists damage and becomes more durable. The longer concrete hardens, the stronger it becomes. Providing the right conditions for concrete to harden is called "curing". Curing depends on the three factors-moisture, temperature and time.

Moisture

Hydration of cement stops when concrete dries. This means that concrete must be kept away from drying for as long as possible to reach its highest strength. There is more than enough water in most fresh concrete to fully hydrate the cement. If the water is kept in concrete, hydration will continue but if the free water evaporates hydration practically stops.

Temperature and time

Warm concrete hardens faster than cold concrete and during the first few days, gains strength faster than cold concrete. In fact, where high strengths are needed quickly concrete may be heated with steam or by other means. When such curing is used, precautions must be taken to avoid damage to the concrete.

At temperatures just above freezing, fresh concrete hardens very slowly and will freeze below 32 F. In cold weather, when temperatures average below 40 F, concrete must be protected against freezing for the first day or two.

When to start curing

The statement "starts curing immediately" cannot be overemphasized. Curing should start as soon as placing and finishing is completed. If concrete hardens, or is alternately wet and dried, cracks may appear on the surface. To prevent such cracks, steps must be taken to reduce evaporation during finishing operations and for several days thereafter. This task may be extremely difficult unless someone has planned and is ready to take action before concrete is cast. All materials and equipment needed for protection from early drying and curing should be available and ready for use before the concrete arrives.

Curing methods

There are many methods of curing concrete. Concrete must be kept continuously moist during curing to avoid crazing and cracking. Choose one method or a combination of methods that is most practical for each job from those listed below.

Wet burlap or mats

Concrete may also be cured by covering it with wet burlap, blankets, cotton mats, or carpets. These materials must be kept wet during the entire curing period. Burlaps should be heavy weight (to hold water better) and should be thoroughly rinsed before it is used. Strips of burlap should be overlapped by about half their width to provide double layer. Burlaps and other absorbent materials can be used on vertical as well as horizontal surfaces. They must be kept wet continuously by frequent sprinkling with a garden hose.

Waterproof paper or plastic film

Waterproof paper made of two layer of Kraft paper bonded together with asphalt and reinforced with fibers is used to cure concrete. It is delivered in rolls but lies flat, when rolled on to concrete pipes. If used carefully, waterproof paper can be used several times. When placed on a pipes, edges, should be lapped and weighted down to avoid wind damage and to be sure that the paper is in contact with concrete.

Plastic film must be at least 4 mils (0.004 in) thick and can be clear, white or black. For hot weather, white is the best, because it reflects the heat; black is better for cold weather because heat is retained. Plastic sheets are not heavy as waterproof paper and sizes tend to be larger. They do not lie fiat; instead they wrinkle. These wrinkles leave a light and dark streaks which might be objectionable. Where appearance is not important, plastic sheets can be used as curing covers. Plastic film or waterproof papers should cover exposed edges as well as flat surfaces.

Curing compounds

Curing compounds are liquids which are applied to the surface of concrete. They form a film which seals the moisture in. To form a continuous film, the liquid must be applied thick enough generally between 150 and 200 Sq. ft per gallons (8 to 10 mils thick). To be effective curing compounds must form a film soon after they are applied. White or gray colours are sometimes added to curing compounds to reflect sunlight and show that the compound is being applied uniformly.

The use of curing compounds is probably the most popular curing methods. Unfortunately it is probably the most abused. Sometimes compounds are thinned, other cases the compound is spread too thin.

Curing compounds are usually applied by spraying the equipment but for small areas they can be applied by brush or roller. Dyes added to curing compounds help the craftsman applying the material to assure uniform coverage.

Formwork

With favorable temperatures, curing continues while metal forms or forms with impermeable faces are in place. This is mainly true for walls and columns which have forms on all sides. Exposed slabs and beams need one of the above means of protection on the top side. If forms are removed before the concrete is strong enough, use one of the other methods for curing until required strength is reached.

Duration of Curing- Hardening of concrete requires time so concrete should be cured for toppings underlayment and other "match patches" used to level floor slabs. Many curing 7 days by preventing loss of moisture whenever the concrete temperature is above 40 F. If the air temperatures are below 40 F, or are expected to fall below 40 F during the curing period, refer to "Cold weather Precautions".

Hot weather precautions

Most concrete is cast in warm weather and warm moist conditions are nearly ideal for concreting.

If the weather is hot, dry, windy a number of steps should be taken to avoid problems with placing concrete in hot weather are:

1. More water needed to make concrete workable.
2. Warm concrete dries fast.
3. Rapid slump loss may occur.
4. Concrete may set too fast.
5. Handling, finishing and curing may be more difficult and may require effort.
6. Plastic shrinkage cracking is more likely to occur because the surface is more likely to dry before curing begun.
7. Concrete is more likely to crack and craze.
8. Strength gain will be slow after 7 days.
9. Concrete contracts more as it cools if its temperature is high.
10. Concrete will be more porous and shrink more because more water must be used in the mix.

Effect of high temperature on concrete

Hot weather affects the properties of fresh concrete in many ways, most of which can lead to trouble unless steps are taken to avoid problems. Concrete tends to stiffen faster in hot weather than at cooler temperatures. For example, if the initial setting time us likely to be reduced to about an hour or less at 95 F. Because of this, the concrete crew should scheduled concrete delivery no faster than the workers and equipments can handle the concrete.

Concrete expands when it is heated and contracts when it is cooled. Concrete made at 95 F will be contract more than concrete made at 70 F when the concrete is cooled to 50 F. If this contraction is resisted by friction, the concrete may tend to crack.

Causes and prevention of plastic shrinkage cracking

Plastic shrinkage cracks form while concrete is soft or plastic, usually soon after concrete is placed. These cracks are most likely to appear when the concrete is warm, the weather is dry (low relative humidity) and windy. Such cracks are located at random and generally appear in Fig. 3.7 (Similar cracks appear in materials other than concrete for example in clay soils or in mud flats.

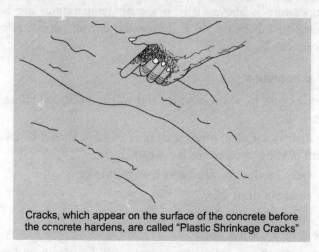

Cracks, which appear on the surface of the concrete before the concrete hardens, are called "Plastic Shrinkage Cracks"

Fig. 3.7

The given figure can be used as a guide to help anticipate when plastic shrinkage cracking might occur. The figure is based on evaporation of water from a flat surface. The chart taken into account the air temperature, relative humidity, wind velocity and concrete temperature. Air temperature relative humidity and wind speed are reported several times a day by most radio stations. Concrete temperature can be taken easily with a thermometer. When the rate of evaporation exceeds about 0.2. Ib of water per sq.ft per hr, plastic shrinkage cracking may occur. With concrete that doesn't bleed much plastic cracking can occur at evaporation rates lower than 0.2.

For example, suppose a weather report says that the temperature of 85 F can be expected and the relative humidity will be about 20% and the wind speed will reach 10 mph. Suppose also that the concrete temperature will be 80 F. To estimate the rate of evaporation:

1. Enter graph in Fig. 3.8 with the air temperature 85 F and draw vertical line to the 20% relative humidity curve.

2. From that point, draw a horizontal line to the right to concrete temperature, 80 F curve.

3. From that point draw a vertical line down to the wind velocity line for 10 mph.

4. From that point draw a horizontal line to the left and read the rate of evaporation from the vertical scale.

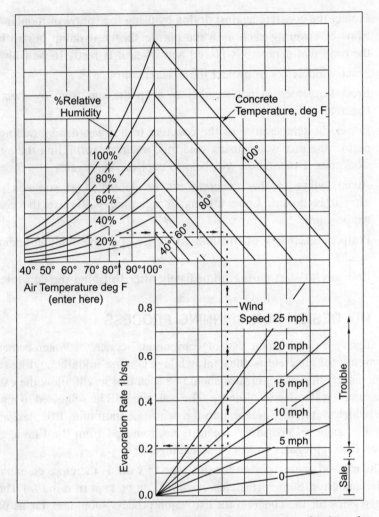

This chart provides a method of estimating the rate of evaporation of moisture from concrete. If the rate of evaporation approaches 0.2 Ib/sq.ft/hr, precautions against plastic shrinkage cracking should be taken

Fig. 3.8

If above these steps correctly, we should get about 0.2 Ib/sq. ft/hr, so we expect plastic cracking to occur unless you take appropriate steps such as those given below. Note that if the wind speed is 15 mph instead of 10, the rate of evaporation will be nearly 0.3 instead of 0.2. Similarly if the concrete is warmer than 80 F, the evaporation rate will be greater.

If plastic cracking might occur, a number of steps can be taken to minimize the problems:

(a) Protect the concrete against drying by using fog sprays or monomolecular films or covering such as white plastic sheets or damp burlap between the time that concrete is placed and when it is ready to be finished.

(b) Erect windbreaks to protect the concrete surface.

(c) Erect sunshades in hot weather to keep the sun from overheating the concrete surface.

(d) Lower the temperature of the concrete in hot weather by adding ice as part of the mix water, using cold mix water, or sprinkling the aggregate stockpiles. Evaporation will help to cool the aggregates

(e) Avoid delays so that concrete can be deposited as soon as possible after it is mixed. Long mixing or agitating will raise the concrete's temperature.

(f) Postpone each step of finishing as long as possible without endangering results.

(g) Be ready to start curing immediately after finishing the concrete.

3.4 MIX DESIGN FOR SPINNING PROCESS

It is slightly different than that of conventional concrete. Water cement ratio in conventional concrete is constant, while it is more initially, and less finally. Combined grading of aggregates should be such that it will allow the extraction of maximum water due to spinning. The concrete will be subjected to a pressure under which water will escape into the centre of spun unit. It is desirable that under the action of pressure, cement is not removed from the fine aggregates nor the fine aggregates from coarse aggregates.

The method of mix design for this type of work is therefore essentially one of trial and error. Some principles, however can be kept in mind (a) The grade of aggregates for the concrete for extraction process should, as far as possible, be continuous, (b) Concrete to be cohesive, not susceptible to segregation when being handled before processing, (c) The proportion of very fine aggregate particles below, say, 600 micron (no. 25) sieve must be sufficient to avoid excessive segregation but on the other hand must not be so high to reduce the rate at which the water can escape during processing. The desired percentage passing on sieve 600 micron (25 No) to be 20 to 25%.

3.4.1 Mix Design Process

It is the adjustment of all the ingredients of concrete in proper proportion so that the mixture will have adequate workability initially and should give the required strength after compaction at 28 days. Concrete is made up of cement, coarse and fine aggregates, water and air. How these ingredients are selected is given on next page.

(i) Cement

For concrete of strength 20 kg/cm^2 the minimum cement content in Indian standard is given as 375 kg/cm^2 and 400 kg for strength of 25 kg/cm^2 (Pipe of dia. 1800 mm and above, EN 630 specification gives minimum cement content as 300 kg/cm^2, other important such as Japan and Australia, minimum cement content is not givens. We will consider a cement of 400 kg/cm^2 for practical purposes.

(ii) Water

For spun concrete, initial w/c ratio is between 0.42 to 0.45, which after spinning expected to reduce to about 0.32.

(iii) Air

Irrespective of compression and cement content this air content cannot be reduce to less than 2.0% of weight of concrete.

(iv) Aggregate

Strength to concrete is gives by aggregates and cement but for that selecting the relative percentage of concrete aggregates to skillful job. For spun concrete after many year of experience, curves have been developed for different mix range of 12.5, 20 mm dia. aggregates. This selection required few trials to arrive at the combined percentage of coarse and fine aggregates. How this is done is shown by mean of actual example is Table 3.1 this is principally based on the sieve analysis of individual aggregates.

From the example it is clear that following percentages of different aggregates gives better combination

CA_1 40%

CA_2 35%

FA_1 25%

The sand grading shown fall in the grading curves given for sand.

3.4.2 Mix Design Example

The mix will normally consist of the following,

1 cm of concrete, with Density of 2450 kg

1 cement 400kg

1 Water w/c = 0.32 is $400 \times 0.32 \dfrac{128}{528}$ kg

Aggregate will be 2450 − 528 = 1922

CA_1 0.40 × 1922 = 768.8 kg i.e. 769 kg

CA_2 0.35 × 1922 = 672.7 kg   673 kg

FA_1 0.25 × 1922 = 480.5 $\dfrac{480}{1922}$ kg

Mix details for 1 cubic meter concrete

Material	Weight kg	Specification	Volume
Cement	400	3.10	0.129
Water	128	1.0	0.128
CA_1	769	2.65	0.290
CA_2	673	2.64	0.255
FA_1	480	2.64	0.182
Air	—	—	0.049
			1.033

Suppose 900 mm dia. 2.43 m long NP_3 class pipe is to be made

Volume of pipe 1.256 m^3

Weight of pipe @ 2450 kg/m^3 3077 kg 3.08 M.T.

Wall Thickness 100 mm.

Materials required will be,

Add 5% for wastage

Cement 502 kg

Water 161

CA_1 965

CA_2 845

FA_1 $\dfrac{603}{3075}$

As the thickness of pipe is more than 70 mm, hence the feeding of concrete should be done in two,

1st layer 70 mm Thick weight.

2nd layer 30 mm Thick weight.

Actual the mix design should be slightly different for both the layer. For 1st layer performable CA_2 and FA_2 shown to be there while for 2nd layer CA_2 and FA_1 and FA_2 should be there, water for the first layer should be with 0.32, however total water should be adjusted.

Therefore actual work of the mix should be,

	Cement	Water	CA_1	CA_2	FA_1	Total
1st layer	351	113	675	592	422	2153
2nd layer	151	48	290	253	181	923
Total	502	161	965	845	603	3076

In this mix 5% should be added for wastage, as said above the mix can further be adjusted having more water with only CA_1 and FA_1 in the 1st layer, so that the mix water will be taken out during spinning of the first layer. In this way a pipe of good quality will be obtained in short time.

3.4.3 Spinning Speeds

For achieving compact concrete, required pressure has to be developed and it depend upon spinning speed. Higher spinning speed is always desired but needs perfect balance of mould. Generally spinning speed with acceleration of 20g to 30g are practicable (g is acceleration due to gravity). During feeding concrete in the mould, the speed of mould to be 3g so that concrete which goes on top during rotation should not fall.

Following formula is used to determine speed of mould during spinning.

Angular acceleration

$$\alpha = V^2/r$$
$$V^2 = \alpha \cdot R$$

For acceleration $\alpha = 3g$ during concrete feeding

$$[2\pi(r)\ N/60]^2 = 3.g.r$$

Where, α = Acceleration

V = Velocity of mould in m/sec.

g = Acceleration due to gravity in m/sec.

N = No of revolutions of mould per minute.

t = Thickness of pipe in meters.

r = Internal radius of pipe in meter.

Therefore,

$$V = 2 \times 3.14 \times r \times N/60 = (3 \times 9.81 \times r)^{0.5}$$
$$0.105 \times r \times N = 5.42 \times r^{0.5}$$

$$N = \frac{5.42 \times r^{0.5}}{0.105 \times r} = \frac{51.7}{r^{0.5}} = \frac{51.7}{0.45^{0.5}} = \frac{51.7}{0.67} = 77.18 \text{ RPM}$$

For acceleration $\alpha = 30g$

$$V = 2 \times 3.14 \times r \times N/60 = (30 \times 9.81 \times r)^{0.5}$$
$$0.105 \times r \times N = 17.16 \times r^{0.5}$$

$$N = \frac{17.16 \times r^{0.5}}{0.105 \times r} = \frac{163.4}{r^{0.5}} = \frac{163.4}{r^{0.5}} = 163.4/0.67 = 243.8 \text{ RPM}$$

For acceleration $\alpha = 40g$

$$V = 2 \times 3.14 \times r \times N/60 = (40 \times 9.81 \times r)^{0.5}$$

$$0.105 \times r \times N = 19.8 \times r^{0.5}$$

$$N = \frac{19.8 \times r^{0.5}}{0.105 \times r} = \frac{188.6}{r^{0.5}} = \frac{188.6}{0.45^{0.5}} = \frac{188.6}{0.67} = 281.5 \text{ RPM}$$

Table 3.2 Spinning speeds in revolution per minute — r.p.m.

DIA.mm	During Feeding	During Spinning	
	3g	20g	30g
250	138	315	390
300	127	290	355
400	116	260	320
450	110	245	300
500	104	235	290
600	95	215	265
700	88	205	250
800	82	190	235
900	78	180	220
1000	74	170	210
1100	70	165	200
1200	67	155	190
1300	64	150	185
1400	62	145	180
1500	60	140	175
1600	58	135	170
1700	56	135	165
1800	55	130	160
1900	53		
2000	52		

Higher speed makes a more compact concrete. However to sustain higher speeds the balance of moulds must be good i.e., both the endrings should be more or less of equal weight. For normal pipe, speed must be 20 g or more.

3.5 ENVIABLE PROPERTIES OF SPUN CONCRETE

Although very little information is available in the literature on the properties of spun concrete, there are several indications and unpublished experiences that allude to improved material properties. Abeles reported some of his experiences in the design and manufacture of centrifugally cast concrete masts and pipes in Europe, and pointed to the improvement gained from the spinning operation. No specific test results were given, but the fact that the spun concrete is much stronger than conventional (non spun) concrete is stressed throughout the paper. Several other studies discussed centrifugal casting, as it relates to the manufacture of pipes and alluded to the advantages of the method. The Prestressed Concrete Institute, The American Society of Civil Engineering and the American Society of Testing Materials, have also realised the importance of this growing industry and published report and standards addressing the design and use of concrete poles with special attention given to the spun poles.

It has been shown by U.S. Bureau of Reclamation that concretes made by spinning are:

(a) About thousand times as impermeable as good (engineering) quality structural concrete placed in situ.

(b) Pipe density of spun concrete is higher than conventional concrete.

(c) **Only concrete of this standard, have the direct tensile strength to resist, without cracking, the primary stresses in pressure pipes. Therefore cracking, inherent in normal reinforced flexural member is impermissible.** The stresses in the concrete pipe wall at test pressure, do not usually exceed 300 to 400 psi. (21 to 28 kg/cm^2) Since cracking is negligible, the stress in steel can be only that in concrete, multiplied by Es/Ec. At test pressure, this might be up to 3000 psi (211 kg/cm^2) or increasing to 15000 psi (1055 kg/cm^2) at concrete failure. This is a small fraction of ultimate tensile strength (about 80000 psi [562 kg/cm^2] to 100000 psi [703 kg/cm^2] of steel. Whatever the amount of reinforcement in the section, having to remain virtually uncracked, the concrete definitely provides 90 to 100% of the tensile resistance. **Thus the traditional concept of cracked concrete is quite inapplicable to pressure pipes.**

Since the steel is doing so little, why do we provide so much? One answer is that, in order to specify the amount, the clause reverts to the orthodox assumption that the concrete is incapable of resisting tension. This is illogical, since the concrete will then crack and releases the pressure which causes the tension. As a device, however, it has the merit of needing no new skill in design to ensure the superior performance under shock load for which R.C.C. pipes are noted.

(d) Their behavior differs radically from that of brittle pipes (Plain Concrete, Cast Iron, Asbestos Cement, Ductile Iron). These subjected to accidental pressure beyond their capacity, burst irrevocably. But an over stress in R.C.C. pipe can crack and leak until it has shed the excessive pressure.

(e) At the instance of cracking, the reinforcement is invoked. It holds the sides of the crack together tightly, or even surrounding moisture ensures autogenously healing. The marble which soon fills the crack is usually stronger than the original concrete. This action is ensured by suitably moderate stress in reinforcement so that it prevents opening of the crack and hence, any tendency to stretch the steel is of negligible dimension, of the crack width.

(f) Criteria for satisfactory shock (water hammer) resistance are well established. Hence the cross-sectional area of hard drawn wire spirals in a length 'L' of pipe with internal diameter 'd' for test pressure 'P' is adopted as $\dfrac{PdL}{40000}$.

These findings, no doubt establish the supremacy and suitability of spun concrete for pipes. **However, as the properties of spun concrete are not known from the standpoint of strength nor permeability, very little information is available in the literatures. Improvements in the physical properties of the spun material do occur, but can not be estimated reliably to incorporate in the design. The situation is further complicated by the unavailability of standard method for casting and testing spun concrete cube.**

Hence a survey was undertaken to know the physical properties of spun concrete. Accurate knowledge of the properties of the material would, ensure a safe and economical design, as well as an efficient utilization of the material.

Findings of the survey undertaken for knowing different properties of spun concrete

The primary goal was to collect information of certain physical and mechanical properties of centrifugally cast (spun) concrete and compare them with the corresponding properties of conventionally (static) cast material. The compressive and tensile strength, modulus of elasticity, permeability and absorption results were searched for. The findings are given below. (Experimental work undertaken by the Development Department of The Indian Hume Pipe Co. Ltd., Mumbai between 1994 – 1996).

3.5.1 Compressive Strength and other Structure Properties of Spun Concrete

Results were available for correlation between spun and vibrated concrete, by three different methods as given below.

(i) By cutting core from the pipe

When different processes other than spinning came into vogue, **American Concrete Pipe Technical Committee** (sub 30 – project 3367) arranged to make, same diameter of pipes, in different factories and conducted tests on pipes made by different processes such as, Tamp, Packer head, wet cast, dry cast and spun. For determination of the compressive strength, cores were cut from these pipes. The average strengths were as given in Table 3.3

Table 3.3 Average compressive strength of concrete made by different processes

Dia. of pipe inches (mm)	Average compressive strength in psi (MPa) at 28 days					Ratio S/V
	Tamp	Packer head	Wet cast	Dry cast V	Spun S	
30" (762)	5226 (36.23)	8830 (60.88)	5211 (35.92)	7075 (48.78)	9940 (68.53)	1.40
36" (914.4)	–	8980 (61.91)	5305 (36.57)	6790 (46.81)	11180 (77.08)	1.65

The compressive strength of concrete at 28 days from spun concrete was more by 1.4 or 1.65 times than that of vibrated concrete or minimum 40%.

(ii) By fixing cube moulds, inside spinning mould (india)

Normal cube moulds were fixed inside a 900 mm diameter pipe mould. The moulds were filled with concrete in the same way as cube moulds, and spun. Vibrated cubes for comparison were also made from the same mix.

The tests were conducted in different factories in India around 1978 to 1990. The results are given below.

Fig. 3.9 Method of casting cubes by spinning

Table 3.4 Results of cubes casted by spinning and vibration process

Ref. No	Density Kg/m³		% increased in density	Spun to vibrated (S/V) strength ratio		
	Spun	Vibrated		1 day	3 days	28 days
F1	2757	2530	2.85	2.01	2.07	1.54
F2	2514	2440	2.95	1.98	1.64	1.40
F3	2514	2452	2.85	1.96	1.47	1.26
F4	2600	2533	2.54	2.12	1.67	–
F5	2594	2519	2.89	2.28	1.86	1.42
F6	2601	2543	2.23	1.94	1.68	1.36
Average	**2565**	**2502**	**3.4**	**2.05**	**1.73**	**1.40**

The correlation factors were different at different ages. They varied from 2.05, 1.73 and 1.4 at the age of 1, 3, and 28 days respectively. The density varied from 2565 to 2502, an increase of about 2.52%.

(iii) From Literature

A graph of water cement ratio against strength is given below, from the book" Properties of Concrete" by A.M. Neville, page, 729.

Table 3.5   Compressive strength of different water cement ratios at different ages

W/C Ratio	Compressive strength in Kg.'cm² at Days			
	1	*3*	*7*	*28*
0.34	215	450	580	740
0.42	85	230	330	490
Ratio of concrete strengths at initial W/C ratio of 0.34 and 0.42	2.5	1.95	1.75	1.5

The graph shows the strengths at different ages of concrete. The ratios of concrete strength for W/C ratio of 0.34 against 0.42 are 2.5, 1.95, 1.75 and 1.5. This confirms more or less the correlation as obtained by fixing cubes in the mould as shown above at 1, 3, 7 and 28 days. W/C ratio of 0.42 may be considered at initial stage and W/C ratio of 0.34 after spinning.

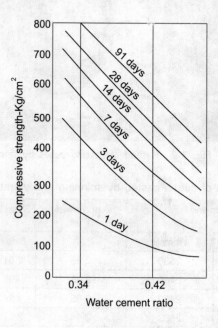

Fig. 3.10   Relation between compressive strength and water cement ratios

3.5.2 Tensile Strength

Tests were conducted on the specimens cut from the pipe and casted one, as per the sketch below.

Average tensile strength of spun concrete specimen was 31 kg/cm² and average tensile strength of vibrated concrete specimen was 20 kg/cm².

3.5.3 Permeability

Testing permeability of concrete has not been generally standardized so that the value of the coefficient of permeability quoted in different publications may not be comparable. There is a further problem with permeability testing namely that, in good quality concrete there is no flow of water through the concrete. For such concrete, test has been given in German Standard DIN 1048. The recommended test age is 28 days. After the sample has been placed in the test cell as shown, a water pressure at 100 KPa (1 bars) is applied for 48 hours followed by pressure of 300 KPa (3 bars) and 700 KPa (7 bars) for 24 hours each. The sample

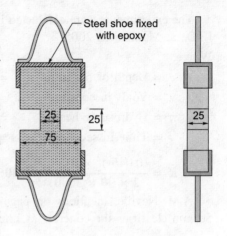

Fig. 3.11   Specimen for tensile tests

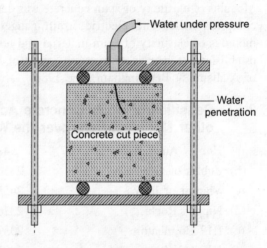

Fig. 3.12   Permeability test set up as DIN1048

is then removed from cell, the surface is dried and the sample is split.

The greatest water penetration depth, measured on tested concrete is taken as an average value of the greatest penetration depth on three test specimens. Tests so conducted, gave average penetration of 46 mm i.e., 0.046 m. The specification recommends that concrete with penetration less than 60 mm, is impermeable.

Coefficient of permeability is calculated as per the formula suggested in book of A.M. Neville on page 495.

$$K = \frac{e^2 v}{2\ ht}$$

The experiments were conducted in the laboratory of ELKEM INDIA PVT. LTD., at Mumbai in 2005.

Where,

e = Depth of penetration = 0.046 m

v = Voids in concrete = 0.04

h = Hydrostatic head = 90 m

t = Time in seconds = 3600 sec

$$K = \frac{(0.046)^2 \times 0.04}{2 \times 90 \times 3600} = 1.306 \times 10^{-10} \text{ m/sec} = 1.306 \times 10^{-8} \text{ cms/sec.}$$

A.M. Neville, mentions on page 495 that depth of penetration less than 50 mm classifies the concrete as impermeable.

3.5.4 Modulus of Elasticity

Modulus of elasticity of spun concrete was determined using both a mechanical compress meter and electric strain gauges. The relationship between the modulus of elasticity of spun material and normally consolidated concrete was established. (Synopsis of research report of Dr. Froud H. Foudd – University of Alabama at Bermingham June 1988)

3.5.5 Densities of Spun Concrete Achieved after Spinning by other Manufacturers over the World

(a) Humes Australia   2541 kg/m^3

(b) Zublin of Germany   2550 kg/m^3

(c) Vianini of Italy   2455 kg/m^3

(d) Nippon Rocla   2326 kg/m^3

(e) IHP Coimbatore   2339 kg/m^3

(f) IHP Miraj   2493 kg/m^3

(g) IHP Hyderabad   2250 kg/m^3

The density also depends upon the specific grade of aggregate.

3.5.6 Conclusions

Following conclusions can be drawn

(1) The compressive strength of centrifugally cast concrete is appreciably higher than that of conventionally cast concrete. The strength co-relation factor is dependent on both mix composition and age of concrete. The factor is about 2.0, 1.73 and 1.4 for 1, 3 and 28 days respectively as compared to conventional concrete.

(2) The effect of spinning is more pronounced in concrete mixes of higher slump.

(3) Compressive strength of core cut at 28 days from spun sections is about 40% more that of cores cut from conventional concrete.

(4) The modulus of elasticity of spun concrete at 28 days of age was about 28% greater than that of conventionally cast concrete.

(5) An increase in density of concrete of about 3.5% on an average did occur, as a result of spinning.

(6) Coefficient of permeability was very small indicating an impermeable material. The ratio of coefficient of permeability of conventionally cast and spun cast concrete was about 3.

(7) A reduction in the absorption of the concrete of about 22% on average, did occur and is attributed mainly to the spinning operations.

(8) Numerous research investigations have demonstrated that the W/C ratio has the largest influence on the permeability of the concrete. It controls the porosity of the paste, which in turn controls the permeability of concrete.

To sum up

The results of the survey demonstrate that centrifugally cast concrete possesses improved mechanical properties and permeability characteristics that surpass those of conventionally cast concrete produced of the same material and cured under the same conditions. Centrifugal compaction induced by spinning, extracts the excess water in the fresh concrete and result in significant reduction in the volume of voids in the hardened material. The influence of voids on the properties of concrete is of paramount importance. The spun concrete process reduces voids and this result in superior strength and permeability characteristics.

3.6 PIPE MAKING FACILITIES

3.6.1 Spinning Machine

Centrifugal spinning machine is the main machine of pipe manufacturing facilities. It consists of revolving wheels (runners) on a shaft mounted on girders. The revolving wheel assembly is connected to a large capacity variable speed mortar, the revolving speed of which can be changed continuously. The mounting girder frame should have high strength, strong enough to take any shocks. The two girders should be connected together by cross girders welded to them. The concrete foundation of the girders should have flexible material as shown in Fig. 3.13 for bigger dia of pipes, the shaft connecting revolving wheels should be strong enough, about 150 mm dia. The runner should be as shown in Fig. 3.13 it should be fitted to the shaft by means of a taper bush and not, by key way and wedge which causes eccentricity. For changing the wearing face of runners, a tier should fitted on the runner as shown in the figure. The wearing surface of tier should be of hard steel to reduce the wear and the time for changing of tier.

The essential part of the machine is series of identical mould specially sized and shaped to produce the particular concrete item required. At the beginning of the operation, a preassembled reinforcing steel framework is positioned in each of the moulds, which are located longitudinally along a common axis. A predetermine amount of premixed concrete is then poured into each of the individual moulds in a way which distributes the concrete evenly along the bottom of the mould.

The entire mould assembly is cradled in two sets of rollers, each set having a common axis which is parallel to the axis of the mould assembly. Each of the roller axes is offset from the vertical plane by an angle of about 35 degrees. All axes are set in a horizontal orientation, longitudinally and transversely. One set of roller are drive roller while the other set are merely idlers. The drive rollers are themselves driven by an electric motor whose shaft is connected longitudinally to the roller shaft. Thus, when the motor is operating, the rollers will be rotated in one direction, say clockwise, which then imparts a rotation of the mould assembly in the opposite direction and results in the rollers, also being rotated clockwise.

To provide the necessary quality of compaction to the concrete contained within, the mould assembly is turned sequentially at three specific speeds, begging with the slowest speed. At predetermine times, the speed is increased to intermediate and finally to the highest speed, and rotated for reestablished periods. Although the various speeds and period may very depending on the particular concrete mix, types of moulds, and model of the spinning machine being used, the actual values of each is based of the following considerations:

1. The lowest speed of the rotation of the mould is all important because it provides for the uniform distribution of the concrete in mould assembly. Without this initial stage, the distribution would be uneven, and the resulting pipe internally eccentric.

2. Although high rotation rates may result in greater compaction, experience has shown than when the speed is to high, components of the mix tend to separate out from each other, according to their specific gravities, and ultimately, the machine itself may be damaged. Optimal compaction speeds have been found to provide compaction forces, between 30 and 40 g. (1g being the usual force of gravity).

3. Smooth, vibration-free rotation of the mould assembly is essential to quality products. Improper rotation or excessive vibration will prevent the

required centrifugal compaction and the final product again may show anomalies.

4. To avoid these adverse affects of vibration two special design features have incorporated into advanced type of spinning machine: first, to ensure that the machine is absolutely rigid, it is mounted an a special base the mass of which is some fifteen times that of the entire mould assembly. Second, each of the rollers is fabricated with a heaver layer of vibration absorbing rubber which will readily compensate for any shock or impact resulting from movement of the load within the rotation mould assembly.

3.6.2 Mould

It is an important component in the manufacturing process, as good pipe will not produced it the mould is not accurse ard properly balanced.

The mould basically consists of a sheet of steel plate with and ring for rotating mould on the machine. The main parts of the mould are:

(a) Mould shaft.

(b) End ring.

(c) Ti rods for connecting shell and end ring.

(d) Socket former.

(e) Crane or electric hoist.

(f) Demoulding frame.

(a) Mould shaft

It is basically made from steel plate by bending it is give the shape of pipe. The inside diameter of the shell is equal to the outside diameter of pipe. At the end of shell this thickness plate of give required shape of spigot or socket by machine and to maintain the shaft. The shell has one or two joint or opening of the mould and taking away the pipe. This is only one longitudinal joint due to daily opening and closing of the mould spigot looser the shape and difficult to correct the shape, while the shape i.e., exact and equal, uniform diameter of socket is possible in the case of protecting socket mould, it is not so for the spigot. Hence two pieces mould properly stiffened is always problem from correct socket and spigot diameter.

At seam joint angle should be welded to the seam plate, then angle should have holes for fixing met end bolts. The most time consuming operation in Demoulding and assembly the mould is fixing and removing the seam bolts. This can be easily a complied by fixing stopper to prevent the removable of the load of the mould so that the seam bolt can be opened and closed by one spanner or fixing a housing for the load of mould as shown in Fig. 3.13.

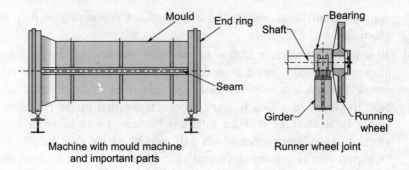

Fig. 3.13 Seam

At tightening the bolts a tape should be fixed on the seam as shown in the figure, at the bolts for the seam fixing should be high tensile steel so that there left is more.

(b) End rings

The play the important parts of rotating the mould smoothly on the runners. Their outside diameter should be inside diameter of pipe. The end ring now should be fabric attend on as shown in the figure. In the case of projected socket mould the weight of the rings (Socket and Spigot) ranges shown preferably is same. This is not always possible hence higher speed on the mould cannot be given.

(c) Tie rods

The function to achieve perfect assembly of mould shell and end rings. In the post long tie rods extending from on end to other were used, but consume lot of time in assembly and stripping. Seam interred short tie rods about 450 – 500 mm long fixed. These should be a bracket on the steel and that an end ring in fastener to the shell.

(d) Socket former

This is an attachment inserted on socket side to form the shape of socket. This is a fabricated from steel and fixed to socket and by means of clips. This clip can be easily turn to enable removing or fixing socket from the socket end rings (see Fig. 3.14)

Hair pin leaver

It is getting suspended from electric hoist. It goes inside the mould and remove the pipe from the mould.

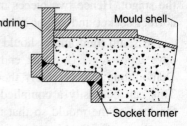

Fig. 3.14

(e) Crane or electric hoist

This is mainly to suspend the mould part during assembling and stripping operation. It has to high the mould with or without pipe and take it is stripping area or to pipe curing area.

(f) Demoulding and assembly frame

It mainly consists of to pipes on which the mould can be place and turned if required.

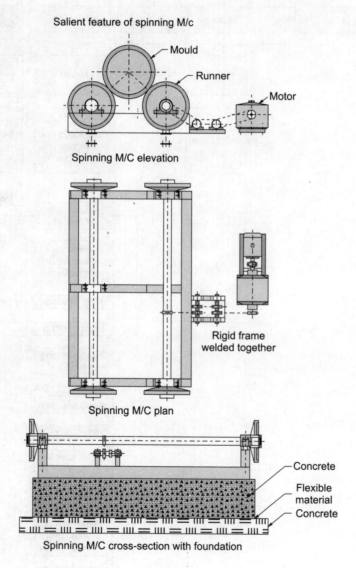

Fig. 3.15

STAGES IN MANUFACTURER OF CONCRETE PIPE

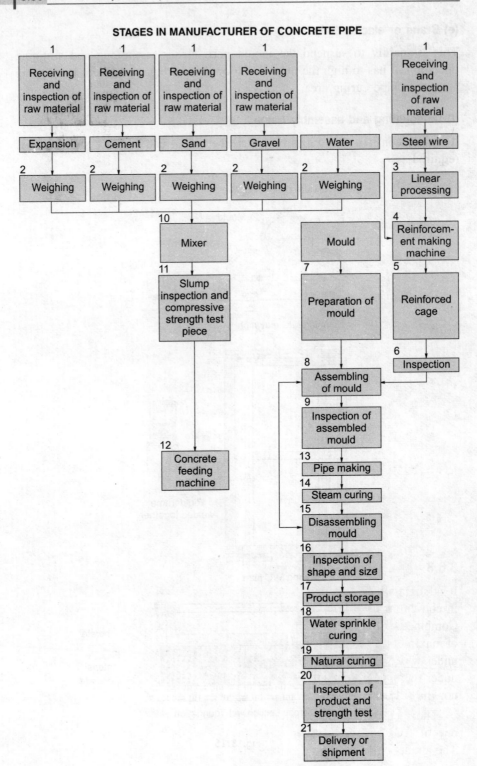

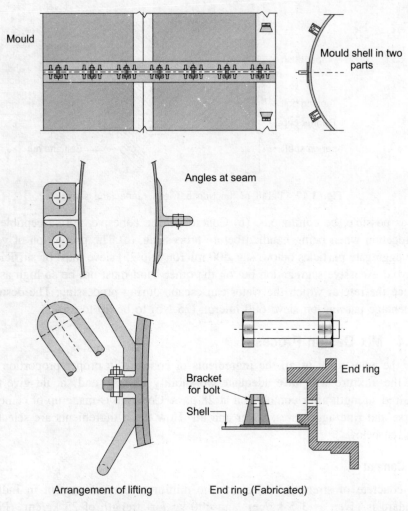

Fig. 3.16 Salient features of mould

3.6.3 Mix Design for Spinning Process

It is slightly different than that of conventional concrete. Water cement ratio in conventional concrete is constant, while it is more initially, and less finally. Combined grading of aggregates should be such that it will allow the extraction of maximum water due to spinning. The concrete will be subjected to a pressure under which water will escape into the centre of spun unit. It is desirable that under the action of pressure, cement is not removing from the fine aggregates nor the fine aggregates from coarse aggregates.

The method of mix design for this type of work is, therefore essentially one of trial and error. Some principles, however can be kept in mind (a). The grade of aggregates for the concrete for extraction process should, as

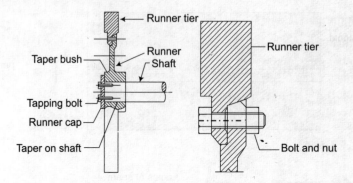

Fig. 3.17 Detail of junction between runner and shaft

far as possible, be continuous, (b) Concrete to be cohesive, not susceptible to segregation when being handled before processing, (c) The proportion of very fine aggregate particles below, say, 600 micron (No 25) sieve must be sufficient to avoid excessive segregation but on the other hand must not be so high as to reduce the rate at which the water can escape during processing. The desired percentage passing on sieve 600 micron (25 No) to be 20 to 25%.

3.6.4 Mix Design Process

It is the adjustment of all the ingredients of concrete in proper proportion so that the mixture will have adequate workability initially and should give the required strength after compaction at 28 days. Concrete is made up of cement, coarse and fine aggregates, water and air. How these ingredients are selected is given below.

(i) Cement

For concrete of strength 20 kg/cm^2 the minimum cement content in Indian standard is given as 375 kg/cm^2 and 400 kg for strength of 25 kg/cm^2 (Pipe of dia. 1800 mm and above, EN 630 specification gives minimum cement content as 300 kg/cm^2, in other important specification such as Japan and Australia, minimum cement content is not givens. We will consider a cement of 400 kg/cm^2 for practical purposes.

(ii) Water

For spun concrete, initial w/c ratio is between 0.42 to 0.45, which after spinning expected to reduce to about 0.32.

(iii) Air

Irrespective of compression and cement content the air content cannot be reduce to less than 2.0% of weight of concrete.

(iv) Aggregate

Strength to concrete is gives by aggregates and cement but for that selecting the relative percentage of coarse aggregates to skillful job. For spun concrete after many year of experience, curves have been developed for different mix range of 12.5, 20 mm max size Aggregates. This selection requires few trials to arrive at the combined percentage of coarse and fine aggregates. How this is done is shown by mean of actual example is to be given Table No 3.6. This is principally base on the sieve analysis of individual aggregates.

From the example, it is clear that following percentages of different aggregates gives better combination

CA_1 40%

CA_2 35%

FA_1 25%

The sand grading shown, fall in the grading curves given for sand.

3.6.5 Mix Design Example

The mix will normally consist of the following, for one cubic meter of concrete

Density 2450 kg

Cement 400 kg

Water $w/c = 0.32$ $400 \times 0.32 = \dfrac{128}{528}$ kg

Aggregate will be $2450 - 528 = 1922$ kg as shown below:

CA_1 $0.40 \times 1922 = 768.8$ kg i.e. 769 kg

CA_2 $0.35 \times 1922 = 672.7$ kg 673 kg

FA_1 $0.25 \times 1922 = 480.5$ $\dfrac{480}{1922}$ kg

Mix details for 1 cubic meter concrete

Material	Weight Kg	Specific Gravity	Volume
M^3	400	3.10	0.129
Cement	128	1.0	0.128
Water	128	1.0	0.128
CA_1	769	2.65	0.290
CA_2	673	2.64	0.255
FA_1	480	2.64	0.182
Air			0.049
			1.033

Suppose 900 mm dia. 2.43 m long NP_3 class pipe is to be made

Volume of pipe 1.256 m^3

Weight of pipe @ 2450 kg/m^3 3077 kg 3.08 M.T.

Wall Thickness 100 mm.

Materials required will be,

Cement	502 kg
Water	161
CA_1	965
CA_2	845
FA_1	$\dfrac{603}{3075}$

As the thickness of pipe is more than 70 mm, hence the feeding of concrete should be done in two, layers.

1st layer 70 mm Thick weight.

2nd layer 30 mm Thick weight.

Actual, the mix design should be slightly different for both the layer. For 1st layer preferable CA_2 and FA_2 should not to be there, while for 2nd layer CA_2 and FA_1 and FA_2 should be there, water for the first layer should be with 0.32, However total water should be adjusted.

Then for actual work the mix should, (for one pipe)

	Cement Kg	Water Kg	CA_1 Kg	CA_2 Kg	FA_1 Kg	Total Kg
1st layer	351	113	675	592	422	2153
2nd layer	151	48	290	253	181	923
Total	**502**	**161**	**965**	**845**	**603**	**3076 kg**

In this mix, 5% should be added for wastage, As said above the mix can further be adjusted having more water with only CA_1 and FA_1 in the 1st layer, so that the max water will be taken out during spinning of the first layer. In this way a pipe of good quality will be obtained in short time.

3.7 JOINTS IN CONCRETE PIPES

A joint is a structure by which two bodies fit together. When two pieces of pipe are joined together, it is possible to produce a joint with zero clearance. The most elemental "zero clearance joint" is one in which two joining ends fit together perfectly with enough pressure to maintain zero clearance. This of course is a bit impractical on anything but a machined joint, so in concrete pipe an annular space is kept, which is caulked either by cement mortar or by inserting an elastic resilient rubber gasket between the two surfaces.

Table 3.6 Aggregate grading of aggregate

Aggregate	%	(Sieve analysis) percentage passing									
		20 mm	16 mm	12.5 mm	10 mm	4.75 mm	2.36 mm	1.18 mm	600 mcr	300 mcr	150 mcr
C.A.1	–	100	68.75	24.67	10.92	–	–	–	–	–	–
C.A.2	–	–	–	96.25	84.12	28.00	9.45	5.87	2.85	1.5	0.68
F.A.1	–	–	–	–	–	99.00	93.00	75.35	52.00	26.10	2.00
F.A.2	–	–	–	–	–	–	–	–	–	–	–

Table 3.7 Mix design-trial 1

	%	20 mm	16 mm	12.5 mm	10 mm	4.75 mm	2.36 mm	1.18 mm	600 mcr	300 mcr	150 mcr
C.A.1	40	40	27.50	9.87	4.37	–	–	–	–	–	–
C.A.2	30	30	30.00	28.88	25.24	8.40	2.93	1.76	0.86	0.45	0.20
F.A.1	30	30	30.00	30.00	30.00	29.70	27.90	22.61	15.60	7.83	0.60
F.A.2	–	–	–	–	–	–	–	–	–	–	–
Total	100	100	87.50	68.74	59.60	38.10	30.83	24.37	16.46	8.28	0.80

Table 3.8 Mix design-trial 2

	%	20 mm	16 mm	12.5 mm	10 mm	4.75 mm	2.36 mm	1.18 mm	600 mcr	300 mcr	150 mcr
C.A.1	40	40	27.50	9.87	4.37	–	–	–	–	–	–
C.A.2	35	35	35.00	33.69	29.44	9.80	3.41	2.05	1.00	0.53	0.24
F.A.1	25	25	25.00	25.00	25.00	24.75	23.25	18.84	13.00	6.53	0.50
F.A.2	–	–	–	–	–	–	–	–	–	–	–
Total	100	100	87.50	68.56	58.81	34.55	26.66	20.89	14.00	7.05	0.74

JOINTS IN CONCRETE PIPES
ENCLOSER - A

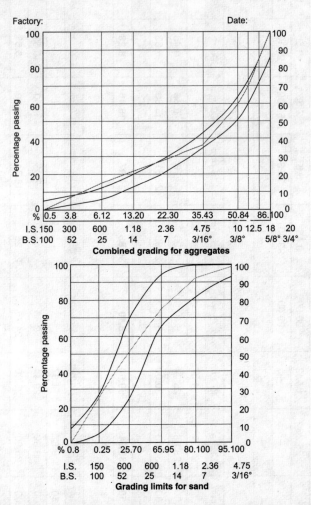

Factory: Date:

Combined grading for aggregates

Grading limits for sand

A water-works engineer would prefer a pipeline without a joint or with minimum joints, but it is not practicable due to restrictions imposed upon the handling equipment, transport facilities, widths of the roads etc.

With pipeline, joints are indispensable. We have only to see how best a joint can be designed, how best it can be made and laid at site and how best it can be maintained to ensure a trouble free service.

One of the most important factors in the development of new pipe material is the joint. Though Freyssinet applied the prestressing technique to produce high pressure concrete pipes as early as 1935, he could not successfully develop a good joint for it. Prestressed concrete pipe, therefore, became a reality only after a suitable joint was developed. The joint in a pipeline is therefore as important, though not more, as the pipe itself.

Environment around a Pipe Joint

Before taking a review of various joints used for concrete pipes, covering their merits and demerits, it is worthwhile to study the environments in which, the concrete pipe has to work. Underground pipeline in the past, was supposed to be subjected to the stresses due to hydrostatic pressure, overburden, defending upon the bedding condition etc. In the joint design, therefore, provision was made only for these considerations which were based on empirical formulae.

Improvements introduced in the past 10-15 years, changed fundamentally, the empirical traditional methods to more rational once, based on scientific principles of structural and soil mechanics. A buried pipeline is now recognized as a load bearing structure. In addition to the normal forces, additional force do come into action in actual service. It is realized that portions of ground are subjected to subsidence even at shallow depths, and also in most alluvial soils. Further, whenever differential settlement is possible, pipelines do not remain undisturbed, but get invariably bent, stretched o shorted by natural forces in soil.

Such forces include vibrations, volume changes due to moisture contents in soil etc. Besides the exposed, as well as underground pipelines, to some extent, are subjected to thermal changes.

3.7.1 Essential Requirements of the Joint

The requirements of a joint for underground pipelines are listed below:

(a) It should preserve the efficiency of the pipeline as a water tight structure throughout its useful life.

(b) It should permit the ground movements to occur without disturbing the joint or causing leakage in the pipeline.

(c) It should withstand external load.

(d) It should be simple, efficient and economical.

(e) It should be easy for maintenance.

(f) Easy for quick assembly at site, - and should ensure a satisfactory joint in reasonable time.

(g) Possibility of jointing under varied conditions.

(h) Should enable individual pipe to be disassembled rapidly without damage to adjacent pipes. Should withstand various conditions due to fluctuating hydraulic flows such as instantaneous high pressures or vacuum created in the pipelines, etc.

(i) Should not allow infiltration i.e., outside water to enter inside.

(j) Capable of providing facility of individual testing of joints as the pipes are laid in stages.

3.7.2 Types of Joints

From the points of views of Structural Engineering, the joints in use for concrete pipes, fall in two categories viz.: (i) Rigid joints and (ii) Flexible joints.

(i) Rigid Joints

The most common type is the collar joint, where the water seal is effected by applying cement mortar which is caulked in the annular space between the collar and the pipe. Compactness and the friction between mortar and concrete surfaces determine the pressure which the joint can withstand.

This joint is like a conservative father, who does not allow any liberties to his children. He maintains that all the family units should follow rigidly family traditions, even in changed circumstances. As all individuals are not alike, so the surroundings. One of the units yields to the influence of changed atmosphere with the result that the entire integrity is lost. Similarly, as the conditions to which different pipes are subjected are not similar, nor the strength of each individual joints, if one of them fails, the entire pipeline fails.

(ii) Flexible joint

A flexible joint is one in which a water seal is effected because of contact pressure between the sealing ring and the pipe surface. The property of rubber to act as a solid flexible spring has been used as seal. The compressive stress induced in rubber results in a force extorted by the rubber against the concrete surfaces. This force creates a unit pressure of rubber against concrete, which holds back hydrostatic pressure equal or less than the said unit pressure. Following are some of the forms in use.

(a) Roll on joint

In this, a ring (circular in cross section) is placed at or near the end of the spigot and rolls along it as the spigot enters the socket. This joint was developed by Cornileous around 1932.

(b) Confined gasket

Rubber ring of circular cross section is held in the groove formed on the spigot. Sometimes, the cross section is in the shape of lip. The lips are opened due to water pressure which ensures water seal.

(c) Sliding gasket

The gasket is fixed on one of the concrete surfaces either socket or spigot. The surfaces are moved relative to each other. The surface are sealed by a number of fins on the head of the gasket which press against the opposing concrete face.

Friction between the ring or gasket and the moving concrete surface during jointing is reduced by use of lubricant, which washes off when the pipe is in service.

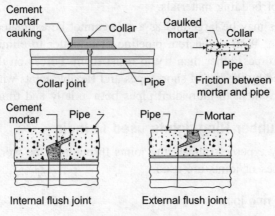

Fig. 3.18 joints in concrete pipes

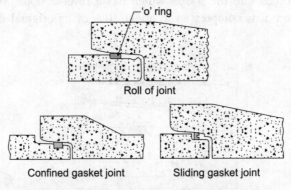

Fig. 3.19 Flexible joints (Commonly used)

Causes of Failures of Pipelines due to Secondary Loads

Fractures of rigid pipe in service have shown that in some cases, as underground pipeline is subject to the simple vertical load and other forces which contribute to the total tensile stress on the pipe materials. Some of the additional causes of stress which commonly occur are given below:

  (i) Settlement of embankments or fills.
 (ii) Settlement of building and structures.
(iii) Restraints caused by manholes.
(iv) Volume changes in clay soils due to wetting and drying.
 (v) Compaction and traffic over-fills.

(vi) Non-uniform bedding or foundation.

(vii) Differential subsidence.

(viii) Vibration caused by machinery or traffic.

(ix) Erosion of bedding materials.

These forces may be large and as a rule cannot be estimated quantitatively with confidence. With a concrete pipeline, it is better to eliminate or reduce the secondary forces rather than try to resist them. Forces on the pipeline can often be reduced by the use of short pipes and flexible joints which allow some freedom of movement to individual pipes both axially and transversely.

Verities of Rubber Ring Joints used in India

There are many types of rubber ring joints throughout the world, but in India mainly two types of joints are used.

(a) Roll on "O" ring joint

In this joint the "O" ring is placed tightly on spigot. It then rolls when the spigot is gradually forced into the socket which has a reverse slope. When it comes to final position, it is compressed to about 40% of its original diameter.

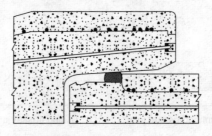

Fig. 3.20 Roll on joint

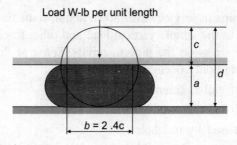

Fig. 3.21 Rubber Ring compressed between two plates

(b) Confined ring joint

The rubber ring is positioned in a groove on the spigot. The ring is tightly fixed in the groove on the spigot and pushed into socket. The ring is compressed and takes the shape as shown. As the ring does not move it is termed as confined ring joint shown in Fig. 3.22.

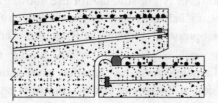

Fig. 3.22 Confined joint

Sealing action of circular chord is accomplished by compressing it between two surfaces as shown in Fig. 3.23. The sealing action is dependent on contact pressure between ring and its mating surfaces.

The "O" ring seal as applied in Concrete Pressure Pipe is a self energizing system where in, the initial ring, pressure resealing from deflection is increased by the pressure of the contained liquid. As a result, the pressure at deflected ring interface is always higher than that of contained liquid.

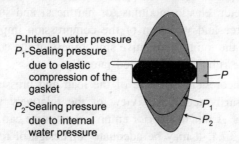

P-Internal water pressure
P_1-Sealing pressure due to elastic compression of the gasket
P_2-Sealing pressure due to internal water pressure

Fig. 3.23 Sealing pressure due to elastic compression of gasket

The essential requirements for satisfactory operation of this type of joint seal are

(a) A combination of rubber hardness and joint clearance that precludes extrusion of the ring from the joint.

(b) Sufficient ring deflection to permit the seal to be self energizing.

(c) Adequate structural strength of socket and spigot to maintain design dimensions under maximum load conditions.

(d) Adequate retaintation of applied stress in the ring chord.

Choice of Rubber for Ring

Natural or synthetic rubber is used for rings. Requirement of rubber are given in I.S. 5382 or B.S. 2494. In this respect it is worth noting that the underground conditions of darkness, wetness and scarcity of ozone both inside and outside of the pipes, favour the preservations of both natural and synthetic rubbers.

Natural rubber however is not suitable for use above ground, when exposed to the air or sunlight: or below ground in the presence of oils, fats or solvent. Synthetic rings of Neoprene, Butadiene, SBR can be used in such cases.

For roll-on joint with a given gap width, harder the rubber, the less its uncompressed diameter must be. Another consideration is the possible differential compression of the ring at crown and invert caused by any tendency to deferential settlement of the jointed pipes, or the transmission of loads from one pipe to another when bedding is not as completely uniform as it should be under these conditions, too soft a rubber might not maintain a watertight seal in some joints. One balance, however and because of its lower stress relaxations, it seems preferable to utilize the softer and larger diameter ring. In current practice, hardness does not exceed 60 shore durometer hardness.

3.7.3 Design of Rubber "O" Ring Joint (Roll-on Type)

Evidently both the initial contact pressure and the radial load imposed by the ring on the pipe socket will depend upon the fluid test pressure to be resisted, the diameter, elastic modulus (or hardness) and stress relaxations of the ring. The degree and variation of the compression imposed on it, by the gap dimensions of the joint, the relative smoothness of the contact surfaces and whether or not, the rubber is restrained laterally.

The rigorous theoretical estimate of the load and pressure is difficult and probably unnecessarily tedious in view of the relatively wide variations in practical conditions. A much superior empirical estimate can be made as shown in given Example 3.7.4. It may be adequate for design of rolling "O" rings in the absences of more exact analysis. It is not applicable to "O" ring restrained in a groove.

The essential principle to be observed in the selection of the size and hardness of rolling rubber ring are

(a) The effective long term contact pressure between the ring and the pipe surface must never be less than the maximum fluid pressure to be resisted.

(b) The frictional resistance of the ring to sliding on the pipe surface must be greater than the force exerted by the maximum fluid pressure on the exposed peripheral projected area of the ring.

To satisfy the first requirement i.e., to resist a maximum long term fluid test pressure of P psi.

The initially mean contact pressure will be

$P' = Fr \; Fc \; p \; lb/\mu^2$

F_r = Factor to allow for stress relaxation

F_c = Compression variation factor

μ = Coefficient of friction between rubber and concrete

p = Long term fluid pressure

To satisfy the second principle, 1 after relaxation of contact pressure, if μ is the coefficient of static friction of rubber and the concrete surface.

Then $2b \; F_c \; p\mu > ap$

3.7.4 Illustrative Example (Roll on "O" Ring Joint) Design of Joint

Data: refer to Fig. 3.20

$d = 20 \text{ mm} = 0.75''$

$a = 0.6 \times 0.75 = 0.45$

$c = 0.75 - 0.45 = 0.30$

$b = 2.4c$

$r = 25\% \text{ - relaxation}$

$$F_r = \frac{100}{(100 - r)} = \frac{100}{75} = 1.33$$

$\mu = 0.7$

$F_c = 1.65$

Then first requirement

$P' = F_r \; F_c \; p = 1.33 \times 1.65 \; p = 2.2 \; p$ i.e., 2.2 times of long term fluid pressure

Second requirement

$2b \; F_c \; p\mu = 2 \times 2.4c \times F_c \times \mu \times p$

$= 2 \times 2.4 \times 0.3 \times 1.65 \times 0.7 \times p$

$= 1.66 \; p$ i.e., 1.66 more than long term fluid pressure

If $p = 15 \text{ kg/cm}^2$ i.e., 213 psi

$P' = 1.66 \times 213 = 354.26 \text{ psi} = 24.9 \text{ kg/cm}^2 > 15 \text{ kg/cm}^2$

Figure 3.21 gives a graph of the sealing pressure for various diameters of rubber rings for different compressions. From the graph for a 20 mm dia. ring, with 40% compressions, the sealing pressure is 25 kg/cm^2 which equal to 24.9 kg/cm^2 pressure as worked above. Actual dimensions of 20 mm dia. ring when used are given Fig. 3.21.

3.7.5 Causes of Failure of Pipeline with Flexible Joints

Though a flexible joint is more reliable and a trouble free, it is not a panacea i.e., cure-all. These joints cannot prevent fractures caused by hard or soft spots in the bedding or in the foundation of pipeline, or by a careless back filling of trench or cracking caused by unduly prolonged exposure of pipeline to hot sun or cold area during construction. It is also necessary to provide proper anchors to the pipeline at all changes of direction.

Compared to collar joint the flexible joint requires proper design, adequate control on manufacture and installation. The rubber ring should be as durable as pipe. The important features of flexible joint include (i) It must provide flexibility and allow the design deflection at the joint, without leakage. (ii) It must transmit shear across the joint without allowing the consequent settlement to cause leakage. (iii) Jointing forces must not be excessive (iv) Rubber ring interface pressures, throughout the life of the line, must be sufficient to prevent root penetration into the line. (v) For water supply and sewerage lines it must be 100% effective in preventing infiltration. (vi) Joint surfaces, spigot and socket, must be manufactured with the necessary low tolerances.

Rubber ring joints in concrete pipe satisfy all these criteria; in fact the joint has been proven as capable of performing at design deflections to pressures well in excess of the capability of the parent concrete pipe. However, if these points are not taken care of, defects as mentioned below may creep in.

Design faults

(a) Lack of effective provision for accurately locating the rolling 'O' ring when making joint. (b) Excessive travel of rolling rings. (c) Excessive or inadequate compression or hardness of rings, or variation on hardness in the same ring. (d) Excessive stretching of the ring.

Manufacturing faults

(a) Variation in the width of the annular gap caused by distortion of the socket or spigot (ovality) or both i.e., by excessive tolerance on variation in dia. (b) (b) Roughness of the contact surface of the socket or spigot. (c) Pipe ends not being true to the axis and (d) Porosity in the pipe.

Rubber ring faults

(a) Rubber of inferior quality or inappropriate physical characteristics. (b) Variation in cross section and or hardness of rubber rings between similar rings or in the same ring and (c) Rubber chord not properly spliced.

Installation faults

(a) Pipe makers instructions not adhered to. (Confined and sliding gasket requires lubrication while roll-on does not).

(b) Rings used after improper storage, e.g., exposed to light and sun over a considerable period causing deterioration of the rubber.

(c) Improper selection of rubber rings.

(d) Rolling rings not placed precisely square to the axis and at correct distance from the spigot end, rings twisted or stretched unevenly, rings damaged by cuts or abrasion, pipe axis not concentric when jointing, spigots and sockets chipped off.

(e) Jointing surfaces not properly cleaned, external annular gap not protected against ingress of bedding or granular material, internal joint gap not cleared of dirt, chips etc.

(f) Tackle used for pulling or pushing during joint making was not of suitable type.

(g) Axial gap between spigot and socket too narrow or too wide.

(h) Non-uniform bedding allowing one pipe to transfer load to the other and so cause uneven compression in the rubber joint ring i.e., by disturbance of bed during jointing.

(i) Sockets supported by hard objects or on foundation soil and thus transferring load of pipe and water to rubber ring which is not designed for it.

(j) Inadequate experience of the pipe layer.

3.7.6 Durability of the Rubber Rings

As long as we compress the rubber and thus cause it to press against the two concrete surfaces, the joint will be water tight. But having produced a seal, what will ensure that the joint will remain satisfactory for 100 year's life expectancy of the concrete pipe? The only answer to this is a rubber material that will retain its elastic physical properties for that period. There is no doubt that rubber can be made to last this long. Rubber ornaments have been dredged from ancient Aztec sacrificial wells estimated to be over 200 years old. Thus under conditions of ideal "Storage" such as we have in a sewer or Aztec sacrificial well where the rubber is moist, cool, away from sunlight and minimum oxygen, it should last.

Doubts are sometimes raised about the durability or rubber rings used for pipes. But there are many illustrations such as Northern outfall sewer near West Ham in London which show that even after a service of 100 years, rubber rings used as flexible gasket have not been deteriorated.

There are many sewers with Roll-on-Joint in service since 1935.

Using rubber rings made of natural rubber is a long establish practice. The physical and chemical requirements of these rings have been established in such documents as B.S. 674-1936. A.S.T.M. C 361 and C-443, AS A 139.

Today use of rubber ring joints for water supply and sewage mains has become almost universal, in all advanced countries. This fact itself speaks for its utility, durability and trouble free service.

3.8 LINING OF P.V.C. SHEET TO CONCRETE PIPES

The material and its function

Plastiline is a plasticized PVC sheet specially designed to be embedded in concrete as a surface protection. Shaped keys provide a mechanical lock with the body of the concrete.

The prime function of plastiline is the protection of concrete against Hydrogen Sulphide (H_2S) attack. It is equally effective against a wide range of acids, alkalis and aggressive salt solutions. Plastiline offers complete protection and is suitable for applications which include sewers, treatment works, industrial waste lines, storage tanks, in fact, any concrete structure where aggressive agents are encountered. Synthetic PVC's high molar mass have proven outstanding resistance to aggressive agencies, both in laboratory tests and under exacting service conditions. The use of a continuous PVC liner to protect sewers and sewerage structures is now backed by many years of service experience in Australia, the USA, Japan, Singapore, Thailand, Bahrain, The United Arab Emirates and Iraq.

Plastiline profile types available

Long Key Plastiline: Designed for use in – vertical cast pipes and in situ structures. This can also be cut into strips of required width for use as water stop.	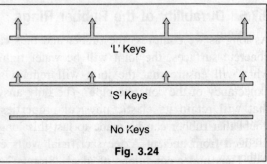 'L' Keys 'S' Keys No Keys
Short Key Plastiline: Designed for use in – spun pipes.	
Plain Plastiline: Designed for use in – bonding applications or situ – actions where no keys are required.	**Fig. 3.24**

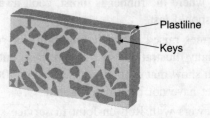

Plastiline

Keys

Fig. 3.25 Section of pipe showing Plastine keys embedded directly into the pipe wall utilization

(a) Principal applications

   a. Lining of spun pipes
   b. Lining vertically cast pipes.
   c. Lining of cast *in situ* pipes.
   d. Water stop at construction joints.
   e. Lining of manholes
   f. Lining in situ structures such as pumping stations and treatment plants.
   g. Pile protection.
   h. Precast concrete component protection.
   i. Basement linings.
   j. Tank lining for chemical manufacturing plant.

(b) Degree of lining required

For minimum cost only those areas of the pipe surface which are likely to be attacked need to be lined. The area needing protection will very with different service conditions. There are basically two conditions:-

(c) H$_2$S Attack conditions only

The sequence of H$_2$S attack-, which finally produces acid, which corrodes the concrete-, occurs only under particular conditions of age, temperature, flow, sulphide content, etc., of the sewage. Only the concrete which is above the minimum flow line will undergo attack, and hence lining and protection is only needed above this flow line. For medium and large sewer pipes the degree of partial lining (see illust.) may only need to be 300° or even less leaving 60° or more of the pipe invert exposed, which is not at risk because it is covered by the flowing sewage. This is particularly applicable to large trunk sewers.

There are situations where the lining must cover must of the pipe surface. These are more common for smaller diameter pipes 375 mm up to 60 mm dia. Although occasionally up to large diameters such as 2100 mm. such lining are conventionally known as 359° linings (see illust.), although this does not mean that the full 359° will be covered. Tolerance variations on pipe diameter and cutting the plastiline to size will mean that there can be a gap at the invert varying from a few millimeter for the small pipes, 50 mm for medium size pipes and 150 mm for large pipes. Even a few millimeters depth of sewage flow will cover any exposed or unlined section.

(d) Industrial Waste Lines

Where such pipelines can carry wastes which are directly aggressive to concrete (either separately or in conjunction with H$_2$S attack) it is recommended that pipes are 360° lined.

In these instances the pipe is first lined with a 359° lining and the resulting unlined invert gap is sealed by one of two alternative methods, (see illust.) so as to completely protect the invert.

Pipes lined with 360° should not normally be buried more than 3 m below the highest level of the water table. This will depend on circumstances and Humes will advise on individual applications.

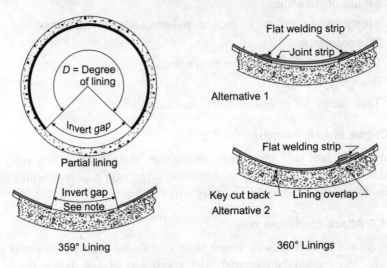

Fig. 3.26

Incorporation in Spun Concrete Pipes

The Plastiline shall be incorporated in the body of the concrete pipe by embedment of the locking keys in the concrete immediately following the manufacture of the pipe.

The plastiline blanket shall be so placed in the pipe that keys are circumferential in direction. This will allow a free escape to the unlined part of the pipe for any moisture or fluid which may accumulate between the lining and the pipe wall.

The extent of the pipe circumference to be covered by the plastiline shall be specified by the client. It is only necessary to line that area of concrete, which is subject to corrosion (see page 3.49).

Immediately upon completion of manufacture of the pipe, the plastiline is to be accurately positioned in the pipe and the locking keys fully embedded in the concrete in such fashion as to develop the pull-out strength stipulated.

Key adjacent to the ends of the pipe barrel shall not be closer than 13 mm to the ends of the barrel and the maximum distance between keys on either side of a pipe joint, with an undeflected pipe fully "home" shall not exceed 130 mm for field-welded pipes. For factory, welded pipes incorporating spigot end caps and/or socket inserts the distance of the first key from the pipe end shall not exceed 130 mm

Embedment of plastiline into concrete pipes can produce minor circumferential corrugation, which will be more evident in smaller diameters. These corrugations shall be in addition to the tolerances allowed and will have no significant effect on the carrying capacity of the pipes because of the smoothness of plastiline and smooth flow conditions.

Incorporation in Vertically Cast Pipes

Plastiline blanket shall be so placed on the inner mould that keys are running in the longitudinal direction. For pipes longer than 2.5 m a circumferential weep channel shall be provided by removing a short length of each locking key to ensure there is no build up of hydrostatic pressure.

The blanket shall be placed under slight tension and held firmly against the inner mould, to prevent leakage of motor between blanket and mould. The recommended method is to use a top and bottom plastic or metal band to hold the blanket in position. Tension in the blanket is obtained by vertical strips at each end, which are fixed with screws from inside the mould.

Normal good practice for vertical casting must be applied, including the following requirements aimed at maintaining blanket continuity:

3.9   3A PROMINENT FEATURES OF MEDIUM SIZE PIPE MAKING PLANT IN JAPAN

Purpose

Main purpose of following information, is to help a new entrepreneur entending to construct a new plant, get over all information, about the plant, so that he can plan his plant considering own requirements.

3.9.1 Typical Concrete Pipe Making Plant in Japan

The plant

Centrifugal reinforced concrete pipe is a kind of concrete pipe, which is manufactured as follows: concrete is compacted by huge centrifugal force (30 – 40 times the acceleration of gravity) of rigid reinforced steel cage rotation as high speed to form the body of the pipe.

This pipe is generally called "HUME PIPE" in Japan and is prescribed in the Japan Industrial Standard as JIS A5303.

The classes of pipe which are actually being manufactured range broadly from small 150 mm inside diameter pipe to extra-large 3,000 mm inside diameter pipe. The standard length of a small diameter pipe (inside diameter 150 mm – 350 mm) is 2,000 mm, and the standard length of pipes with larger inside diameter is 2,430 mm. Depending on requirements, long pipe about 4,000 mm could be made. The following points are features of this method of producing concrete pipes.

1. The concrete body is compacted tightly. There are no bubbles, and so water absorption is small.

2. Surplus water is removed by centrifugal forces; the water: cement ratio of concrete will be small; too, is very smooth; accordingly, resistance to water flow is small.

3. The outside appearance is more beautiful than any other method of production. The inside surface, too, is very smooth; accordingly, resistance to water flow is small.

4. The manufacturing facilities are rather simple and trouble-free. Manufacturing does not required high technical skill.

5. Adjustment of machinery is not necessary for a certain range of inside diameter, so occasional change-over of inside diameter will not lower production efficiency.

The compressive strength of concrete manufactured in this way is more than 400 kg/cm^2 High-class blending will easily produce concrete with compressive strength of about 600 kg/cm^2 varies ranks of pipe body strength could be had by combining concrete strength and amount of steel reinforcement. Recently, by using special blending material, it has become possible to greatly increase the strength against external pressure and internal pressure.

All sorts of pipe joints can be made by selecting the appropriate mould. The socket type using a collar or rubber gasket for the joint (see Fig. 3.1) and the caulking type are employed for the excavation method of laying pipes. Working efficiency of the former is inferior, so the letter type of joint is generally used. The right type of joint, however, should be used, case by case, depending on the objective of using the R.C.C. pipe and the ground condition of the laying site. The standard joint of pipe, which is to be laid by jacking method, is a combination of steel collar and rubber gasket, (see Fig. 3.2) and the body of the pipe itself is thicker than that which is used in the excavation method.

The main uses of concrete pipe are as follows:

1. Sewer system (rain, sewage, drainage).

2. Water works (service water), industrial water conduit.

3. Agricultural water work and water supply.

4. Cross channel duct for freeway.

5. Cable duct.

6. Well wall (The product can also be used as huge foundation material by employing the post-tensioned method of prestress construction).

Process Description

The manufacturing process of this method of production is shown in the attached flow sheet. A simple description according to order of processing will be made concerning the main manufacturing facilities.

1. Construction of the reinforced steel cage

The reinforced steel cage of a R.C.C. pipe consists of circumferential spiral steel wire reinforcement and a longitudinal steel wire reinforcement. By setting up an automatic reinforcement forming machine, the point of intersection of the circumferential spiral reinforcement and the longitudinal reinforcement can be spot-welded quickly and the cage can be constructed in a short time. The longitudinal reinforcement is cut to the designated length by a high speed linear cutter, and is processed directly from steel wire coil.

2. Manufacturing of concrete

In manufacturing concrete, a mixing plant is necessary. The capacity of the installation is decided by the capacity of the hume pipe production capacity. The plant under discussion has a storing bin, weighting equipment, and a mixer. A central control system makes one-man control possible. Material storing tank, cement silo, and blending material tank are ancillary equipment, and these are joined to the plant by belt conveyor, pipe, etc.

3. Concrete conveying system

Concrete is conveyed from the mixer to the various pipe making machines by remote control, and the bucket has an automatic opening – closing lid.

4. Pipe making facilities

The centrifugal machine is the main machine of the pipe making facilities. The revolving wheel is connected to a large capacity variable speed motor, the revolving speed of which can be changed continuously.

Concrete conveyed to the hopper by the concrete conveyor is supplied to the mould by a screw conveyor or a belt conveyor, depending on the diameter of pipe to the manufactured.

5. Steam curing

In order to speed up disassembling of moulds of completely moulded pipe, the mould frame is carried into the sealed steam tank: steam is fed to quicken curing. The capacity of the boiler for supplying steam is decided by the production capacity.

6. Disassembling and assembling of mould

The complete moulded product, which has been completely cured by steam, is placed on the disassembling rack: the mould is disassembled, and the completely moulded product is taken out. After the product is taken out, the mould is transferred onto the mould assembling rack; after being washed, the mould is assembled to take shape as if reinforced steel wire cage was in it.

7. Movement of the mould with pipe

Movement of the mould from the process to another – pipe making → steam curing → disassembling and assembling of mould → completed product – is done mainly by a traveling overhead crane, but chain conveyor is also used for short distance movement.

8. Product storage

The product which has been taken out of the mould is carried by an overhead crane or a forklift to the stock yard, arranged according to diameter and classification, and kept in stock until time of delivery or shipment. During this waiting time, if necessary, wet curing is done by a sprinkler.

Table 3.9   Production capacity

Size of pipe	Range of diameter	Quantity produced in one run	Quantity produced per day
Small diameter	150 – 350 mm	8 – 12 pieces	80 – 120 pieces
Medium diameter	400 – 1,000 mm	3 – 5 pieces	20 – 40 pieces
Large diameter	1,100 – 2,000 mm	2 – 3 pieces	12 – 20 pieces
Extra-large diameter	2,000 – 3,000 mm	1 – 2 pieces	6 – 12 pieces

General Plan for Establishing Plant

As mentioned previously, the diameter of pipe to be manufactured charges from small diameter pipe to extra-diameter pipe; therefore, actual designing of the plant should be based on the scope of pipe diameter required.

Consideration has been given here for manufacturing all sizes of pipes. Accordingly the lineup will be four series; small diameter, medium diameter, large diameter, extra-large diameter. However, depending on the planned amount of manufacturing, each series may be used plurally by changing the range of diameter.

For the purpose of reference, in case of four series, a table will be given which shows the standard inside diameter of each series and the amount of production in one run. The total production in tons is about 30 tons.

Table 3.10 Required machinery and equipment

Item	No of Sets
1. Batcher plant (32 m³/hr)	1
2. Cement conveying equipment	1
3. Cage material conveying equipment	1
4. Concrete conveying equipment	1
5. Steel rod drawing machine	1
6. Reinforcement cage making machine (one for each series)	4
7. Overhead crane (two for each series)	8
8. Centrifugal force, pipe making machine (one for each series)	4
9. Mould assembling and disassembling equipment (one for each series)	4
10. Mould disengaging agent spraying equipment. (one for each series)	4
11. Boiler (steam volume 2.0 tons/hr, pressure 7 kg/cm²	1
12. Compressor (37 kW)	1
13. Pump for water supply	1
14. Draining pump	1
15. Equipment for testing material and finished products	1
16. Tools	1
17. Instruments for measuring and testing	1
18. Maintained and control equipment	1
19. Spare parts for machinery	1
20. Mould (150 mm – 3,000 mm)	174

Table 3.11 Required raw materials per day (including loss)

Item	Quantity
1. Cement	54,800 kg
2. Sand	78,600 kg
3. Aggregate	115,000 kg
4. Mixing material	380 kg
5. Steel reinforcement	5,700 kg

Table 3.12 Required utility

1. Consumption of electric power	1,500 kWh/day
2. Crude oil	700 liters/day
3. Water	40 tons/day

Table 3.13 Required manpower

Item	Nos.
Engineer (Technical)	3
Skilled worker	40
General worker	60
Total	103

Table 3.14 Required area

Main Building	3200 m^2
Other Building	1300 m^2
Land	38000 m^2

Table 3.15 Dimensions of steel collars

Unit: mm

Nominal Diameter	Perimeter	D_e	H	T_{el}	T_e^2	g	Weight Kg
600	2287	719	45	4.5	4.5	3.5	28
700	2664	839	55	4.5	4.5	3.5	33
800	3010	949	60	4.5	4.5	3.5	38
900	3387	1069	70	4.5	4.5	3.5	44
1000	3764	1189	80	4.5	4.5	3.5	50
1100	4109	1299	85	4.5	4.5	3.5	55
1200	4486	1419	95	4.5	4.5	3.5	61
1350	5020	1586	103	6.0	6.0	5.0	94
1500	5586	1766	118	6.0	6.0	5.0	108
1650	6120	1936	128	6.0	6.0	5.0	121
1800	6654	2106	138	6.0	6.0	5.0	134
2000	7376	2336	153	6.0	6.0	5.0	153
2200	8099	2566	168	6.0	6.0	5.0	174
2400	8815	2788	179	9.0	6.0	5.0	256
2600	9538	3018	194	9.0	6.0	5.0	283
2800	10260	3248	209	9.0	6.0	5.0	311
3000	10983	3478	224	9.0	6.0	5.0	340

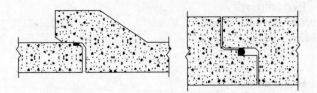

Fig. 3.27 Joint for normal joint

Joint Used for jacking and other pipes were of flexible rubber ring, their sketch is shown below

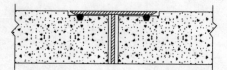

Fig. 3.28   Joint for jacking pipe

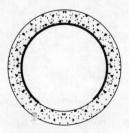

Fig. 3.29   PVC lining

CHAPTER 4

Manufacturing of Welded Cages and Details of Machines

4.1 INTRODUCTION

Presently the cages are made manually or on batten type of spot welding machines, in most of cases, As there are number of arms to which battens are fitted the speed of the drum cannot be increased, so also the socket cage cannot be made together, As such, the production of the machine in low. Fixing, clamping and releasing of the longitudinal is a time consuming operation.

New machine which are developed, avoid all these time consuming operations, with spot welding device. The cage forming machine automatically produce reinforcing steel cage, by welding special steel to longitudinals at the intersecting points. One can save labour and raw materials, by using these machines, and produce reliable product, not only straight cage but also cage for socket can automatically and continuously be formed by these machines, Further improvement as indicated below are also possible.

(a) The machine automatically forms steel cage, with socket and cuts them in between sockets.

(b) Welding pitch of spiral wire can be varied in the straight and socket parts.

(c) Cage can be formed continuously and automatically.

(d) Height efficiency of production.

(e) No skilled labours is necessary as every operation is controlled by push button and the cage can be made automatically.

(f) The machines are compact in size, can easily change the diameter automatically, of socket and straight portion.

(g) The cage made by these machines are very accurate in diameter, free from bent, wrap, with accurate pitch of spiral sheet.

These are different models depending upon the diameter of the cage. All these, including productively are given in the following pages on next page.

Designation	Dia. Range	Length of Cage
1. 4-1/3.5	100-350 mm	2500 mm
2. 4-5/12	500-1200 mm	2500 mm
3. 4-13/18	1300-1800 mm	2500 mm
4. 4-14/30	1400-3300 mm	2500 mm

4.2   PRINCIPALS OF WORKING

The reinforcing steel cage of a concrete pipe consists of circumferential spiral steel wire reinforcement and longitudinal steel wire reinforcement. By setting up an automatic reinforcement forming machine, the point of intersection of the circumferential spiral reinforcement and the longitudinal reinforcing can be spot welded quickly and cage can be constructed in short line. The longitudinal reinforcement is cut to the design length by high speed linear cutter and as processed directly from steel wire coil shown in Fig. 4.1.

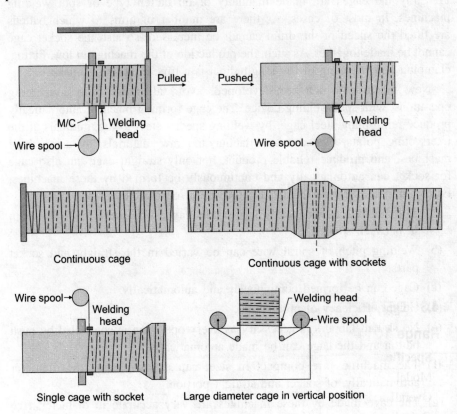

Continuous cage

Continuous cage with sockets

Single cage with socket

Large diameter cage in vertical position

Fig. 4.1

Type 4 - 1/3.5 Diameter Range 100 - 350

Photograph 4

Type 4-4/12 Range 400 - 1200

Photograph 5

4.3 DETAILS OF TYPE 4 - 1/3.5 DIAMETER MACHINE

Range 100 to 350 mm Diameter

Specification

Model	:	4-1/3.5
Overall length	:	7 Meters
Overall width	:	3 Meters
Weight	:	2.5 M.T.
Drum ring running speed	:	Max. 60 RPM

Require power	:	For running ring - 5 kW
		For Welding - 30 KVA
Forming range	:	100-350 mm dia. socket and spigot
Max. length of cage	:	2500 mm
No of longitudinal wire	:	10
Dia. of longitudinal wire	:	2.5 mm to 3.25 mm
Max. carbon content	:	0.2%
Dia. of spiral wire	:	2.5 mm to 3.25 mm
Winding pitch of spiral wire	:	25 mm to 100 mm

Type 4-5/12 Range 500-1200

Photograph 6

Photograph 7

4.4 TYPE 4-5/12

Range 500 to 1200 mm Diameter

This machine is suitable for forming the cages required for pipes of diameter 500 to 1200 mm. The length of the cage is 2.5 mtr. The special feature of the machine is that it will make the socket cage together with the barrel cage of the pipe in one operation.

This machine is highly productive as compared to the batten type spot welding machine of Hume or similar machine of Rocla. In the Hume's machine, there is a drum on which wire is wound. As there are number of arm to which battens are fitted the speed of the drum cannot be increased. As such, the productivity is low. The Rocla machine is similar to this machine, but longitudinal which are clamped on a wheel, actuated by a screw, are pulled as the welding proceeds. When the cage is completed, the longitudinal need to be released from the clamps. Clamping and releasing longitudinal is time consuming operation with this machine.

In the new machine this operation of pulling is avoiding by providing an arrangement of pushing the longitudinal as the circumferential wire is welded. For forming socket the diameter of the barrel is increased automatically as the cage is formed. All the controls for this machine are electronic and one man can change simply by operating a switch.

SPECIFICATIONS

(a) **Overall Dimensions**

Type	4-5/12
Overall length	8000 mm (including cage cradle)
Overall height	2500 mm
Overall width	3800 mm
Total weight	4500 mm

(b) **Forming Capacity**

Range	Straight pipe 500 mm – 1200 mm
	Socket pipes 500 mm – 1200 mm
Maximum length of pipe	2500 mm
No of longitudinal	12 Nos
Dia. of longitudinal wire	4 mm – 6 mm
Dia. of spiral wire	4 mm – 6 mm
Spiral wire pitches	25 mm – 100 mm

Quality of wire	The carbon content not to exceed 0.2% and wire should be without rust.
No of Machine revolution	8 to 15 RPM

4.4.1 Electrical Equipment

(a) Transformer

Welding transformer	AC 415V, Three phase
Rated capacity	30 KVA
Working factor	10 – 15%
Tap change over type	
Water cooling type	

(b) Motors

For the machine rotation	3.7 kW
For returning the ring	1.5 kW
For bender (with brake)	0.75 kW
For moving electrode with trolley	0.4 kW
For changing winding pitch	0.7 kW

(c) Operation Condition

Working place	Indoor
Temperature	Normal temperature
Operation	Continuous
Dust contents	Medium
Daily output	

The daily output depend on the following elements:

1.0 Cage dimensions and length

2.0 Pitch

3.0 Wire dia.

4.0 No of longitudinals

5.0 Frequency of dimension changing

6.0 Wire quality i.e., weld ability and elasticity.

7.0 Nature of wire i.e., degree of dirtiness and corrosion as well as surface condition of wire.

Under normal conditions, the machine is welding the following number of points per minutes:

Wire 4/4 mm	...120 Points/minutes
6/6 mm	... 96 Points/minutes

Following table show the output of finished cages

1200 mm dia. - do -   ...25 Nos/Shifts
900 mm dia. - do -   ... 30 Nos/Shifts
600 mm dia. - do -   ...35 Nos/Shifts

4.5 TYPE 4 - 13/18
Range 1300 to 1800 mm Diameter

This machine is suitable for forming the cages required for pipes of diameter 1300 to 1800 mm. The length of the cage is 2.5 mtr. The special feature of the machine is that it will make the socket cage together with the barrel cage of the pipe in one operation.

SPECIFICATION

(a) Overall Dimensions

Type	4-13/18
Overall length	8000 mm (including cage cradle)
Overall height	3700 mm
Overall width	4500 mm
Total weight	5800 MT
Shipping volume	110 m^3

(b) Forming Capacity

Range	Straight pipe 1300 mm – 1800 mm diameter
Maximum length of pipe	2500 mm
No of longitudinal	24 Nos
Dia. of longitudinal wire	5 mm – 7 mm
Dia. of spiral wire	5 mm – 7 mm
Spiral wire pitches	30 mm – 100 mm
Quality of wire	The carbon content not to exceed 0.2% and wire should be without rust
No of Machine revolution	6 to 12 RPM

(c) Electrical Equipment

(d) Transformer

Welding transformer	AC 415V, Three phase
Rated capacity	60 KVA
Working factor	10 – 15%
Tap change over type	
Water cooling type	

(e) Motors

For the machine rotation	7.5 kW
For returning the ring	3 kW
For bender (with brake)	0.75 kW
For moving electrode with trolley	1 kW
For changing winding pitch	1 kW

(f) Operation Condition

Working place	Indoor
Temperature	Normal temperature
Operation	Continuous
Dust contents	Medium

(g) Daily Output

The daily output depend on the following elements:

1. Cage dimensions and length
2. Pitch
3. Wire dia
4. No of longitudinals
5. Frequency of dimension changing
6. Wire quality i.e., weld ability and elasticity.
7. Nature of wire i.e., degree of dirtiness and corrosion as well as surface of wire.

Under normal condition, the machine is welding the following number of points per minutes:

Wire	5/5 mm	... 96 Points/minutes
	7/7 mm	... 72 Points/minutes

Following table show the output of finished cages

1800 mm dia. NP2 class pipe with socket	... 12 No/Shifts
1400 mm dia. NP2 class pipe with socket	... 16 No/Shifts
1300 mm dia. NP2 class pipe with socket	... 20 No/Shifts

4.6 JUMBO SIZE

Range 1400 to 3300 mm Diameter

Table 4.1

Range	
Product cage diameter	1,400 – 3,300 mm
Product cage length	2,500 mm
H 3000 – 250	

Contd...

Number of axial steel	30
Diameter of axial steel	5 – 7 mm
Diameter of spiral steel	5 – 7 mm
Winding pitch of spiral steel	25 – 100 mm
Number of electrode revolution	8 RPM
Electric power for motors	8.5 kW
Transformer	
Rated capacity	87.0 KVA
Maximum capacity	127.0 KVA
Working rate	23.7%
Control system	SCR con. System
Cooling system	Water Cooling
Dimensions of machine	
Length	7,700 mm
Width	6,800 mm
Height	
H 3000 – 250	4,200 mm
Weight of machine	8,500 kg

Vertical formation of cages prevents the cages from warping which tends to occur when forming Jumbo size cages Horizontally.

Production Capacity: 15 min. for one cage production

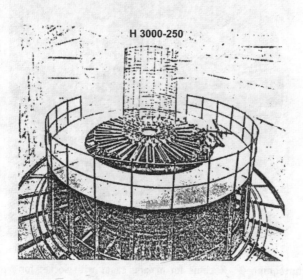

Fig. 4.2

Photograph 8 Machine for making cages with socket for pipes up to 350 mm dia

Photograph 9 Machine for making cages with socket for pipes above 350 mm dia

4.7 CYCLE TIME OF STEEL CAGE PRODUCTION

Table 4.2

Work Item Description	Item No	Cage Production Machine										
		No 1 (Ultra Large)				No 2 (Large)				No 3 (Medium)		
		2250 φ		2500 φ		1400 φ		2000 φ		400 φ		1200 φ
		Inside	Outside	Inside	Outside	Inside	Outside	Inside	Outside	Single	Inside	Outside
1. Inserting of longitudinal Reinforcing steel	(1)	4'30"	4'30"	5'30"	5'30"	2'30"	2'30"	3'30"	3'30"	1'30"	2'30"	2'30"
2. Setting of spiral Reinforcing steel	(2)	2'	2'	2'	2'	1'	1'	1'	1'	1'	1'	1'
3. Welding	(3), (4), (5)	6'	4'30"	16'	12'	3'	2'	5'	4'	2'	5'	4'
4. Taking away Completed cage	(6)	2'	2'	2'	2'	1'	1'	1'	1'	1'	1'	1'
5. Carriage	(7)	1'30"	1'30"	1'30"	1'30"	1'	1'	1'	1'	1'	1'	1'
Total Time		16'	14'30"	27'	23'	8'30"	7'30"	11'30"	10'30"	6'30"	10'30"	9'30"
		30' – 50'				16' – 22'				6' – 22'		

CHAPTER 5

Specifications and Quality Assurance

WHAT IS SPECIFICATION?

The specification gives the principle design required details of raw materials to be used. The process to be used and the acceptance tests.

5.1 DIFFERENT TYPES OF SPECIFICATIONS

5.1.1 Performance

Where the pipe must pass tests, to make positively sure, that it is fit, to do the job for which it is intended.

5.1.2 Prescription

Where the purchaser designs the pipe himself for the field load, nominating sizes, reinforcement details, concrete mix details, manufacturing procedures, etc., and provides supervision of manufacture to ensure that, his wishes are carried out. By providing the design, the Purchaser takes the responsibility, and hopes that his theories have produced a pipe, which will not fail under field load.

5.1.3 Design

Where the purchaser nominates permissible design stresses in the belief that a pipe designed within such limits will not fail under field loads.

Prescription and Design specifications are used for buildings and structures which are difficult or impossible to test. They are of necessarily conservative.

Performance specifications are used for products which can be tested. The purchaser has the assurance that the product will perform as desired in practice, and that the design is economical.

The performance type is used for concrete pipe in Australia, Britain, South Africa and New Zealand, India etc. The prescription type is used in North America, though it should be noted that the designs were based on the results of performance tests.

The Australia standard for precast concrete drainage pipes is AS 31 - 1973. It is quite stringent, compared with overseas standards. It requires that the pipes pass an ultimate test load equal to 1.5 times the cracking Test load, which compares with 1.25 in Britain and 1.25 for class V pipes in North America. It requires that the maximum crack width at cracking load be 0.15 mm (0.006") compared with 0.010″ in Britain and North America. Its requirement regarding absorption are also strict compared with overseas standards.

The fact that the ultimate load is required to be 1.5 times the cracking load means that the pipe has a safety factor of 1.5 against cracking.

5.2   DEVELOPMENT OF SPECIFICATIONS

5.2.1   Development of a new ASTM Specification for "Reinforced Concrete Culvert, Storm Drain and Sever Pipe"

(a) The purpose of this was to summaries the activities of ASTM committee 6–13, Concrete pipe, in development of a new reinforced concrete pipe specification which would replace ASTM C76. The main point of contention has been a proposal to reduce the specified minimum cement content to 470 Ib. (278 kg/m^3) per cu. yd. from 564 Ib. (333.3 kg/m^3) per cu. yd. Without providing other safeguards to insure water tightness and durability.

The existing C76 specification sets forth minimum wall thickness, minimum concrete strengths and minimum steel areas which must be complied with, for each diameter and strength classification.

(b) A minimum cement content of 564 Ib. (6 bags) per cubic yard 350 kg/m^3 and a maximum absorption of 8% was specified. Acceptable curing procedures are briefly described.

(c) For the past six or seven years the American concrete pipe Association has been trying to develop a "performance type" specification for reinforced concrete pipe and culvert pipe. Early efforts were based on including an alternate basis of acceptance in the existing ASTM C76 specification. For various reasons, these efforts were not successful then further efforts have concentrated on developing a separate specification, which would eventually replace the existing C 76 specification.

(d) A draft of a such specification was presented to ASTM committee C-13, concrete pipe, at its meeting in November 1968.

(e) The proposed specification, which has since been called an acceptance specification, was intended to give the pipe producer more latitude, in producing pipe to meet required test loads.

(f) The acceptance specification sets forth standard diameters and strength classes. Except for a minimum concrete strength of 280 kg/cm^2, the structural design is left to the pipe producer. Acceptance of pipe is based either on 3 – edge bearing test results or on designs which can be shown by means of test data to be satisfactory. The minimum cement content is set at 278 kg (5 bags) per cubic meter and the absorption test has been dropped. No curing procedures are described and curing is called for only indirectly. During the following 12 months, John Hendrickson, PCA representative on ASTM committee C-13, although in favor of a performance type specification, consistently opposed the reduced minimum cement content for the following reason:

   (i) Strength alone is not necessarily an indication of durability. Porous concrete can be strong but vulnerable to an aggressive soil condition.

   (ii) Reduced cement content reduces concrete resistance to sulfate attack and perhaps to acid attack.

   (iii) Reduced cement content may reduce the ability of the concrete to protect reinforcement from corrosion, particularly in porous concrete.

   (iv) No data had been provided to show the effect of the reduced cement content on such properties as strength, permeability of concrete pipe.

   (v) The cost saving is very slight in terms of cents per foot of pipe. As a result of this action and support by a few other members, the proposed minimum cement content of 350 kg/m^3. (6 bags) per cu. yd. was reinstated and the acceptance specification was approved at a meeting of ASTM C-13, November 20, 1969 and has been submitted to letter ballot of the full committee.

   (vi) In his letter of December 1, 1969 to members of ACPA, Mr. R. E. Barnes, Managing Directing, briefly outlined the controversy and requested that they seek opinions of their local cement companies "regarding the apparent requirement that precast concrete pipe requires a higher strength and cement content than cast – in – place concrete pipe". Requests for further information have been received from the following member companies, American, Ideal, Kaiser and Lone star. The above reference to "Cast – in – place pipe" refers to the fact that carl Wilder, Manager of the PCA Public works and transportation section, acting as spokesman for ACT committee 346, presenter a proposed standard, "Specification for cast- in –Place Non-Reinforced concrete pipe", for approval at the ACT convention, November 6, 1969.

By way of explanation, there are major differences in the "Concrete properties" portions of the ASTM specification provided only for a 3 'Edge bearing test to determine acceptability, that the concrete strength be 4000 psi (280 kg/cm^2), and that the concrete contain at least 470 lb (280 kg/cm^2) of cement per cu. meter. The ACI specification requires that the 28 days compressive strength of the concrete shall be 3000 psi (210 kg/cm^2 that have a water ratio of not more than 0.53 by weight, and that the concrete shall contain at least 470 Ib. (5 bags) of cement per cu. yd.

Reference to ACI – 631 "Recommended Practice for selecting Proportions for concrete", will show that the cement content will be 280 kg/m^3.

Another, but quite important, difference between the two specifications, is that ACI 346 provides for a hydrostatic test, if such is required by the purchaser. Requiring such a test is quite common when cast-in-place concrete pipe is to be used under even moderate heads.

(viii) Throughout the deliberations, John Hendrickson, the PCA representative, has contended that the proposed reduction in cement content should be supported by experimental data to verify the quality of the pipe or that other safeguard should be provided. He acted in sincere and ethical manner, and not attempted to delay or block the vote on the specification. His professional background indicates he is highly qualified to serve as a member of the ASTM committee. He was on the staff of the American concrete pipe Association for fifteen year before joining the PCA in 1965.

5.3   QUALITY ASSURANCE

5.3.1   Factors which Influence the Quality of Pipes

We have investigation factors which govern the successful manufacture of large pipes especially those with shells thicker than standard. Belfield in particular, have done a considerable amount of work, but other branches have also contributed their own ideas.

Features which influence our thinking on how to make good pipes are:

(a) Inside and outside finish.

(b) Breaking up of grids.

(c) Pipes falling in.

(d) Slumping.

(e) Longitudinal cracks at a very early age.

(f) Longitudinal cracks due to handling at an early age but later than (5)

(g) Longitudinal cracks at an advanced age due to handling, shrinkage etc.

Those are dealt with each of these in turn:

Outside finish

Provide the mix and consolidation are satisfactory no porous patches will be evident, and if an outside finish similar to spun pipes is required, lightly hosing the mould with water before starting to fill will give the required result.

Belfield followed that honeycomb patches were caused to a considerable extent, by a close pitched grid preventing the concrete outside the grid from being consolidated. This effect has been largely eliminated by using a close pitch alternating with a wider pitch. Welders which weld two wires simultaneously can produce this type of grid, i.e., 'S' welder and Belfield's mandrel welder.

Inside finish

Satisfactory mix and efficient filling are necessary. After filling, lightly spray with water and throw in a mixture of sand and cement. Surface is brushed or toweled after removal from machine.

Breaking of grid

Apparently this result is mainly due to four causes:

(i) Weld not good enough. Weld more equal sections together, i.e., 0 gauge is better than 0 gauge to 5 gauge. Grid machine must be working well.

(ii) Longitudinal bars too light. Deformation of longitudinal bar between studs tend to break welds (elliptical grids especially).

(iii) Inefficient studs on elliptical grid break off or bend. Belfield used a tree legged stud. Springvale suggest that insufficient concrete between the grid and the roller contribute to the grid damage and tend to get 20 mm cover. Belfield and Brisbane can manage with 10 mm cover and we this should be possible everywhere.

(iv) Grid jumping dowel at lifting hole. Eliminate by tying grid in mould to prevent turning.

Pipes falling

This trouble is associated with slumping but can be contributed to by goods loose in mould and/or studs collapse and thus hold the grid firmly in position will cutout this effect. Falling in, is more liable to occur in pipes with thicker shells.

Slumping

Associated with falling in and is worse with thicker shells. Measures which will help to reduce this effort are:

(i) Wetter mix.

(ii) Use of calcium chloride

(iii) Roll for at least 10 minutes after filling.

(iv) Increased rolled speed may help.

Longitudinal crack at a early age

These occur in the manufacture and have been practically eliminated by increased speed of the roller. Increase speed has also allegedly improved the outside finish. Springvale have found that for larger pipes increase speed has been a help, but there have been some disadvantages with smaller pipes.

Eliminating of these has been helped by:-

(1) Using high early strength cement (not readily available at all factories).

(2) Taming pipes before stripping.

(3) Use of calcium chloride.

Longitudinal cracks at an advanced age

Belfield and Brisbane apparently do not suffer to any extent, this may be due to local condition such as moist climate, or that pipes are laid soon after making. Inspection of a risbane pipes revealed few cracks even in older pipes.

Springvale use a grid, double at sides and single at top and bottom for all classes of pipe of about 2000 mm diameter and larger. They contend that extra expense is warranted as cracks which occur during storage have been practically eliminated. Whether this practice is necessary is doubtfull, but the drier climate and the longer period before using may have more effect than we think. A double grid may be needed, because available wire would give too close a pitch, and in addition double grid do contribute to the load bearing capacity of the pipe.

Mixing

Mix efficiency test

With the introduction of vertical casting water cement ratio is low. Hence for conductive trail on mix design. The mix should be through, for getting correct results. The mixer should be good, one of such mixture in shown below. A trial of this mixture for mix efficiency should be conducted as shown on next page.

Photograph 10 Laboratory mixer

Table 5.1

Mix. Details	Per Batch Wt.in Kg
1 CEMENT 20%	130
2 CA 2 55%	350
3 FA 1 25%	160
4 W/C 0.42	55 Lit.
TOTAL	**695 kg**

1. Sample taken – 2.0 kg of concrete by weight from actual mix.

W/c ratio is 0.42

Cement content of 20% for 2 kg concrete is

$$\frac{130}{695} \times 2 = \textbf{374 gms.}$$

CA 2 and FA 1 of 80% for 2 kg concrete is

$$\frac{510}{695} \times 2 = \textbf{1467 gms.}$$

Water content for 2 kg concrete is

$$\frac{55 \times 2}{695} = \textbf{158 gms.}$$

2 kg of concrete immersed in water and Analysed. Sample kept for drying in oven.

After drying weight of cement (actual)   = **357 gms.**

After drying weight of CA 2 + FA 1 (actual)   = **1402 gms.**

Percentage difference in size

$$\text{Cement} = \frac{-17 \times 100}{374} = -4.5\%$$

$$\text{CA 2 + FA 1} = \frac{65 \times 100}{1467} = -4.4\%$$

Mix efficiency as per IS IS 1791 – 1968

Material	Allowable % variation	Actual Variation
Cement	8%	4.5
Fine aggregate	6%	4.4
Coarser aggregate	6%	

5.4 MAIN TESTS TO BE DONE

5.4.1 Cube Testing

One of the important acceptance test as cube. Following information will help to cast and trial cubes perfectly.

Preparing the machine for test	
After leaving sufficient space of cubes.	
(1) **Wipe** the **Upper Platen** and **Check** that it is **Clean** (bend down to see if necessary) and then **Wipe** the **Lower Platen.** See Third diagrams, Preparing the cube for test.	 **Photograph 11**
(2) Check the reference number and date to make Sure it is the **Right Cube** to be tested. **Wipe** the surface to remove grit, water and scum. The cube should still be damp. **Remove Fins** carefully with edge of trowel, or with a file, and wipe again to remove grit. **Check Size** with rule or template. Observe whether **Dimensions** are measurable **Incorrect** (particularly the height), or cube out of square. Observe **Condition** of cube.	Curing condition: **Wet Damp** or **Air-Dry** Type of concrete, if special (i.e. no-fines) 7300 15/10/14 Reference mark on cube date of casting. Date of test age at test days. Nominal size: 100 mm, 150 mm (4 in. or 6 in.)

Weighing the cube	
(3) **Weigh** the cube **In Air**, while it is still damp. Place the **Cube in the carrier** in the water. Note for supervisor This may be useful as a check in the event of an inconsistent test result. If the cube is not to be tested immediately, return it to the tank or cover it with damp Hessian untie required for test. This may occur if several cubes are weighted at one time before testing. Wipe the surface again when the cube is to be tested. This does not apply to no – fines cubes.	Information to be entered on CUBE TEST RECORD SHEET Weight in air (kg) = weight A. NB. Weight to be recorded to the nearest 2 grams.

Placing the cube in the machine

Are **Sub – Platen** (or **Auxiliary Platens**) to be used?

No without sub-platen

Place Cube on Machine platen on its.

Side – note trowel led surface at front or on one side. Check that fins have been removed and full contact is made.

Position Cube centrally on guide. Marks on machine platen.

Operate Machine to **Close Space** above cube and apply a small load.

For the **First Cube** of a batch to be tested, remove the load and

Check that the **Load Pointer** returns to **Zero**.

Re-apply a Small Load.

Check No Gap between cube, sub-platens and machine platens- look for daylight.

Crushing the cube

Fix the **Safety Guard** in position.

Operate the machine controls to **Start Loading** the Cube, without shock.

Control the machine to **Apply Load** so that load pointer follows the pacer (on some machines This will be automatically controlled)

Observe the way in which the **Cube** behaves. During the test and note anything unusual (e.g. Cracking on one side of the cube only).

Continue to apply load until the load pointer fall away and continues to do so, taking care to **Maintain Correct Loading Rate** right until failure.

Make sure that there is **Signs of Failure** on the Cube

Observe the **Maximum Load** reached (if there is a Maximum load pointer this should stop at the Maximum load and not fall back).

Information to be entered on **Cube Test Record Sheet:**

Maximum load(kN or tonne force) Compressive strength (MN/m^2) (N/mm^2)

Note for supervisor: Calculate compressive strength by dividing maximum load by nominal cross-section Area express to nearest 0.5 MN/m^2 (N/mm^2)

Release the load gently.

Remove Cube from machine carefully.

Return the maximum load pointer to **Zero.**

Summary

Apply the load at the **Correct Rate**.

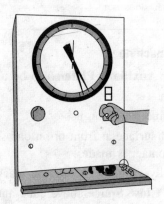

Photograph 12

Until the cube **Fails.**

Note Maximum Load.

Proceed to 7, Examination of the cube after test

Examination of the cube after test.

Examine the crushed cube after failure:

If the appearance is as shown in one of the diagrams of **Normal Failure** (below), record this, No further examination of the cube is required.

If the cube fails in some other way record **Unusual Failure** on the record sheet: Do **not** throw the cube away; keep it, so that your supervisor can look at it to see the type of fracture.

Information to be entered on Cube Test Record Sheet:

Type of failure:

(Record **Normal** failure or **Unusual** failure and note type by reference to illustrations)

Clear Away Debris, remove and clean sub-platens (If used), and wipe the platens.

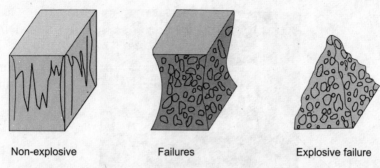

Non-explosive   Failures   Explosive failure

Usual Failures: Equal cracking of all four exposed faces: generally little damage to
faces touching machine platens.

Fig. 5.1 Normal failure

Note for supervisor: Some of the unusual ways in which cubes may fail are
show in the diagrams below.

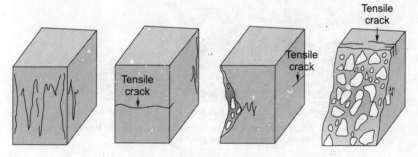

Fig. 5.2 Unusual failure

5.4.2 Checking Socket and Spigot dimensions

Correct Diameter of Socket and Spigot are very important for proper functioning
of Rubber Ring position. Detail procedure for this is given in Chapter 6
under Section 6.4.

5.4.3 Three Edge Bearing Test

5.4.4 Arresting Growth of Algae in Curing Tank

Growth of algae in water of maturing tanks is always a problem. It makes the
water dirty and creates unhygienic conditions.

The main reason for the growth is sun rays. For preventing the growth, a
solution of sodium Hypo chloride (Nao Cl) 55% shall be added to the water in
the tank at the rate of 10 milligrams per liter. This will not cause any harms
to the human beings working in tank.

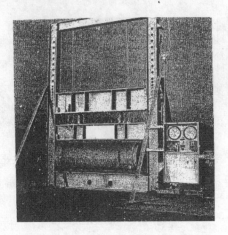

Photograph 13 Method of tests external load test

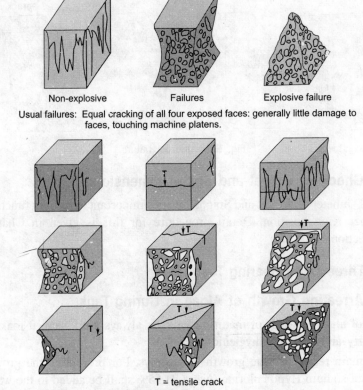

Non-explosive   Failures   Explosive failure

Usual failures: Equal cracking of all four exposed faces: generally little damage to faces, touching machine platens.

T = tensile crack

Usual failures: Excessive cracking of one face or corner. Sometimes accompanied by tensile cracks in one or more faces.

Fig. 5.3 Chart to be displayed near cube testing machine

5.4.5   Socket check Report

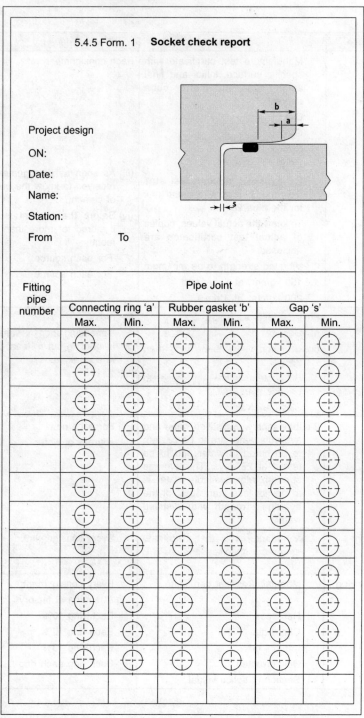

5.4.5 Form. 1   Socket check report

Project design

ON:

Date:

Name:

Station:

From   To

Fitting pipe number	Pipe Joint					
	Connecting ring 'a'		Rubber gasket 'b'		Gap 's'	
	Max.	Min.	Max.	Min.	Max.	Min.

5.4.6   Form for Control Tests for Raw Materials

Table 5.2

Material	Control Test	Frequency
1. Cement	Manufacture test certificate with specific surface, initial and final setting time and mortar cube strength $\left.\begin{array}{l} C_3A \\ C_3S \\ C_2S \\ CFAF \end{array}\right\}$ % of these see 7.4 Compressive strength test at 1 day and 3 days with usual mix for the pipe 28. To know the actual values, copies of actual test certificates are enclosed. Vibrated strength to be minimum 150 kg/cm^2 at 1 day and 200 kg/cm^2 at 3 days.	Each consignment (iii) As soon as consignment is received to know the quality of cement. (iv) Before the cement isused if stored for more than one month. For each source For each source
1. Coarse and fine aggregates	Physical tests such as (i) Density (ii) Crushing value to be less than 25% Silt contents in fine Aggregates to be less than 2%. Sieve analysis	Everyday Depending upon the variation in grading but at least once in a week.
2. Water	3 cubes each with distilled water and available water are to be casted, to be cured in water and to be tested at 28 days. The average strength with available water is to be not less than 80%, of the average strength with distilled water.	For each new sources of water.
4. H.T. wire	Manufactures is test certificate as per our specifications, for H.T. Wire.	Every consignment.
5. Rubber ring	Manufactures test certificate. Hardness Chord dia. Internal dia. Uninterrupted Section at splice visual	Every consignment. At source at factory Each ring 10% Each ring 10% Each ring 10% Each ring each ring

5.4.9 Cement Test Certificates in Following Pages

ACW-'UltrTech 53'

Issue Date: 31.05.2013

ADITYA BIRLA

UltraTech

UltraTech Cement
TEST CERTIFICATE
FOR
53 Grade Ordinary Portland Cement
Test & Compliance

IS 12269

CM/L-205841

Particulars CHEMICAL REQUIREMENTS	Test Results	Requirements of IS: 12269-1987	
1. $\dfrac{CaO-0.7SO_3}{2.8\ SiO_2 + 1.2Al_2O_3 + 0.65\ Fe_2O_3}$	0.92	0.80 1.02	Min Max
2. Al_2O_3/Fe_2O_3	1.32	0.66	Min
3. Insolube Residue (% by mass)	1.19	3.00	Max
4. Magnesia (% by mass)	2.19	6.00	Max
5. Sulphuric Anhydride (% by mass)	1.53	3.00	Max
6. Total Loss on Ignition (% by mass)	1.67	4.00	Max
7. Total Chlorides (% by mass)	0.010	0.10	Max
PHYSICAL REQUIREMENTS			
1. Fineness (m^2/kg)	3.15	225	Min
2. Standard Consistency (%)	28.0		
3. Setting Time (minutes)			
a. Initial	155	30	Min
b. Final	240	600	Max
4. Soundness			
a. Le-Chat Expansion (mm)	1.0	10.0	Max
b. Autoclave Expansion (%)	0.088	0.8	Max
5. Compressive Strength (MPa)			
a. 72 +/- 1 hr. (3 days)	42.0	27	Min
b. 168 +/- 2 hr. (7 days)	51.5	37	Min
c. 672 +/- 2 hr. (28 days)	64.0	53	Min
6. Performance Improver (%)		5.0	Max
a. Limestone	2.0		
b. Fly Ash	NA		
c. Granulated Slag	NA		

The above cement complies with the requirements of
IS: 12269-1987 for 53 Grade Oridinary Portland Cement.

Date of Dispatch: 16.04.2013
Week no: 16 of 2013

HOD(QC)

UltraTech Cement Limited
Works: Awarpur Cement Works, Awarpur, Taluka-Korpana, District-Chandrapur, Maharashtra-442 917

ACW-'UltrTech 53'

Issue Date: 31.05.2013

ADITYA BIRLA

UltraTech

UltraTech Cement
TEST CERTIFICATE
FOR
53 Grade Ordinary Portlant Cement
(Fly Ash based)
Test & Compliance

IS 1489

(Part)
CM/L-1280946

Particulars CHEMICAL REQUIREMENTS	Test Results	Requirements of IS: 1489-1991(Part 1)	
1. Insolube Material (% by mass)	22.39	$X + \dfrac{4.0(100-X)}{100}$ (x = Declared % of Fly Ash)	Max
2. Magnesia (% by mass)	1.98	6.00	Max
3. Sulphuric Anhydride (% by mass)	1.63	3.00	Max
4. Loss on Ignition (% by mass)	1.06	5.00	Max
5. Total Chlorides (% by mass)	0.011	0.10	Max
PHYSICAL REQUIREMENTS			
1. Fineness (m²/kg)	341	300	Min
2. Standard Consistency (%)	31.5		
3. Setting Time (minutes)			
a. Initial	250	30	Min
b. Final	335	600	Max
4. Soundness			
a. Le-Chat Expansion (mm)	1.0	10.0	Max
b. Autoclave Expansion (%)	0.040	0.8	Max
5. Compressive Strength (MPa)			
a. 72 +/– 1 hr. (3 days)	30.0	16	Min
b. 168 +/– 2 hr. (7 days)	40.0	22	Min
c. 672 +/– 2 hr. (28 days)	54.0	33	Min
6. Drying Shrinkage (%)	UT	0.15	Max
7. % of Fly Ash addition	27.0	15.0	Min
		35.0	Max

The above cement complies with the requirements of
**IS: 1489-1991 (Part 1) for Portland Pozzolana Cement
(Fly Ash based)**

Date of Dispatch: 16.04.2013
Week no: 16 of 2013

HOD(QC)

UltraTech Cement Limited
Works: Awarpur Cement Works, Awarpur, Taluka-Korpana, District-Chandrapur, Maharashtra-442 917

Normal Problems Encountered During Manufacture of Socket Spigot Concrete Pipes and their Probable Solutions

General

Problems are usually encountered during manufacture of any product; however, as in manufacture of pipes there are many stages, the problems are little more. The problem may be due to defective equipment, process, mix design etc. For the sake of simplicity, the problems are considered under three different headings.

6.1 PROBLEMS DUE TO MOULD

Non setting of the pipe even after spinning for a long time

There may be a many reasons for this, but the principal reason is non uniform motion to the mould during spinning. This is also called bumping of the mould. The non uniform motion disturbs the setting or stiffening of hardened concrete and makes it plastic like rubber.

Non setting may be, either due to defective mould or machinery. If it is only for a particular mould, the probable reasons are;

(i) Deformation of end ring

Check diameter of end ring at different points as shown. If the diameters are differing more than 1 mm, then a cut has to be taken to the on end ring surface.

(ii) Diametrical difference between two end rings as shown in Fig. 6.2.

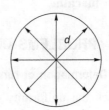

Fig. 6.1 Mould end ring

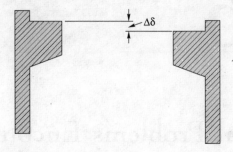

Fig. 6.2 Diametrical difference between two end rings to be less than 1 mm

(iii) Partial unevenness on end ring surface

Unevenness of end ring surface may be due to convexity or chipped off portion due to unknown force during demoulding such as, by hitting hammer on end ring surface. Correct it by taking a cut on end ring.

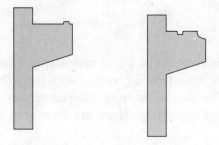

Fig. 6.3 Cross section of end ring

(iv) Deformation and lack of weight balance of mould

Deformation on shape of mould may be caused due to some external forces or by mis-fabrication of mould.

Lack of weight balance may be due to wrong design of mould or too much slurry sticking to the mould.

Important dimensions of the mould which must be checked are given in enclosure.

When bumping appears in all moulds, then the fault will probably be with machine.

6.2 PROBLEMS DUE TO MACHINE

(i) Deformation of runner wheel

Check the uniformity of runner face. This can be done by using deflection meter. Fix the dial gauge as shown in sketch, rotate the runner slowly and record the deflection at every stage as shown in the table.

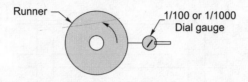

Fig. 6.4 Method of checking deflection on circumference of runner

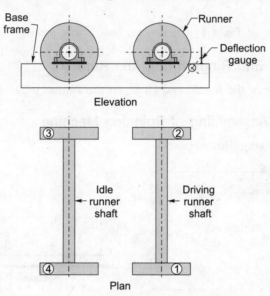

Fig. 6.5 Views of spinning machine: shafts and runners

Readings taken on one machine are shown in table below.

Table 6.1

Runner Number	Circumference of Runner	Diameter of Runner	Deflection Reading	Permissible
	mm	mm	mm	mm
1	1230	391.71	0.10	0.3
2	1230	391.71	0.06	0.3
3	1230	391.71	0.05	0.3
4	1230	391.71	0.05	0.3

As the deflection is less than permissible, runner is O.K.

6.2.1 Level of Runners

Difference in level between two runners fixed on same shaft should not be more than 0.5 mm.

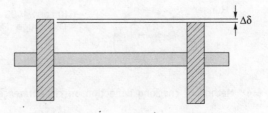

Fig. 6.6 Checking top level of two runners

6.2.2 Mould End Rings Hitting the Runner

Clearance between the mould end ring and the runner is insufficient.

6.2.3 Right Assembling of Spinning Machine

Check the following dimensions.

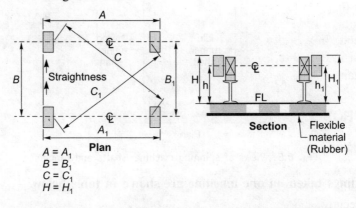

Fig. 6.7 Checking the important dimensions of spinning machine

6.2.4 Non Concentric Motion of Runner

At high spinning speed, this is much pronounced. The runner should be fitted on the shaft by means of taper bush as shown in Fig. 6.8 and not by a key way and wedge. The correct fitting is shown in Fig. 6.8. For changing the runner face, a tire should be fitted on the runner as shown in the Fig. 6.8. Note the clearance between taper and runner.

Even after checking all the above the problem persists, it may be due to foundation concrete below the machine. Some flexible material like rubber sheet below foundation concrete, should be used to absorb the vibrations. Cross girders should be welded to the main girders on which runners are fixed. These girders should then be embedded into the concrete with long bolts. This is essential for production of good pipe. The weight of concrete foundation to be 3 times the weight of moving mass (mould + pipe).

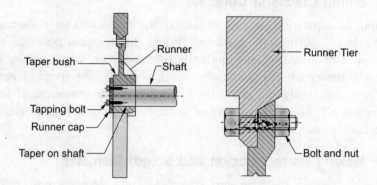

Fig. 6.8 Connection between shaft and taper bush and connection between taper bush and runner tier ring

The diameter of shaft on which the runners are fitted should be minimum 150 mm. PSC pipe mould with projecting socket exert unequal load on runners, when end rings are at both ends; hence shaft must be strong to take shear. For best results, the runners should preferably be of old railway wheels, or of a manganese casting.

6.3 PROBLEMS DUE TO IMPROPER PROCESS FOLLOWED

Demoulding

Before removing the end rings, longitudinals are detensioned. Sometimes longitudinal cracks appear inside core. This is mainly due to unequal tension in consecutive longitudinals especially when core thickness is less or unequal cover on longitudinals.

6.3.1 Partially Filled Socket

If feeding is not done properly especially at socket, air entrapped between socket former and the mould gets on locked up. It prevents concrete, to completely fill the socket, have feeding to be done slowly.

6.3.2 White Patches on the Outer Surface of Pipe Seen after Demoulding

The white patches are the indication of weak concrete. During spinning, if the mix is not properly graded, some water remains in the concrete between the aggregates, instead of coming in the centre. The concrete from where water has not come out has more water i.e., W/C ratio is high. Therefore, it is weak. This happens generally in winter season when viscosity of water is increased and water becomes little thicker and make it difficult to escape.

6.3.3 Slump Cracks in Concrete

Slumping is separation of concrete from the moulds. This is because of concrete is not sufficiently hardened. When this happens concrete cracks. If the speed of mould while spinning is increased fast, water coming out of concrete is prevented. This resulted in weak concrete. The speed of mould is to be increased slowly and gradually within one to three minutes. If the mix contains more fines, the possibility of water coming out is less. In such case, reduce fines in mix.

6.3.4 Making Correct Socket and Spigot Diameter

In spinning process the diameters of socket and spigot are not correct, to achieve correct dimensions, method shown be adopted.

Method of Checking Spigot and Socket Diameter

Spigot diameter

The gauge is as show in Fig. 6.9 shall be used.

Suppose the theoretical outside diameter of spigot = 1200 mm

Permission variation   = ±1 mm

The maximum diameter   = 1201 mm

The minimum diameter   = 1199 mm

To enable gauge to pass over the spigot surface, the diameter must be more by 1mm than the maximum spigot diameter.

The clearance between gauge points   = 1200 + 1 + 1

                                          = 1202 mm

For checking spigot, the gauge is held in the position as show in Fig. 6.10. then by keeping point 'X' fixed, other end 'Y' of gauge is moved over the circumference of spigot, where the gap between spigot surface and gauge is maximum, a strip 2.5 mm, thick and 25 mm wide (straight side) (1201 – 1199 = 2.0 mm) is inserted, it should not go. Two checks shall be done at 90° to each other.

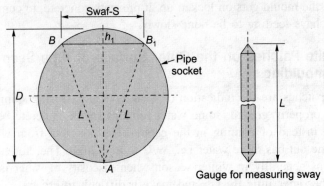

Fig. 6.9 Arrangement for measuring sway

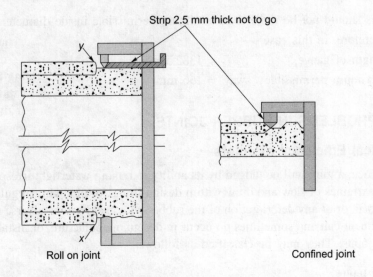

Strip 2.5 mm thick not to go

Roll on joint   Confined joint

Fig. 6.10 Arrangement for checking spigot diameter

6.3.5 Socket and Spigot Diameters

Socket diameter (Jointing surface)

The spigot diameter shall be checked by measuring the sway by touching the two points B and B_1 along the circumference of socket (see Fig. 6.9): BB_1 givens the sway.

Example:

Suppose the theoretical diameter of socket

$$D = 1383 \text{ mm}$$

Permissible variation $\quad\quad = \pm 2$ mm

$$= - 0.5 \text{ mm}$$

The maximum diameter $= D_1 = 1385$ mm

The minimum diameter $= D_2 = 1382.5$ mm

The sway $\quad\quad\quad\quad\quad = S = 2 \times [h \times (D_1 - h)]^{0.5}$

Assume h_1 $\quad\quad\quad\quad = 5$ mm (see Fig. 6.9)

Sway $\quad\quad\quad\quad\quad\quad = 2 \times [5 \times (1385 - 5)]^{0.5}$

$$= 2 \times [5 \times (1385 - 5)]^{0.5}$$

$$= 2 \times 83$$

$$= 166 \text{ mm}$$

Length of gauge $L^2 = (S/2)^2 + (D^1 + h)^2$

$$= (166/2)^2 + (1385 - 5)^2$$

$$= 6889 + 1904400$$

$$L = 1382.5 \text{ mm}$$

This should not be more than minimum permissible inside diameter.
Therefore, in this case

Length of gauge, $L = 1382.5$ mm

Maximum permissible   sway $= 166$ mm

6.4 PROBLEMS IN RUBBER JOINTS

Practical Efficiency of Joints

In practice, a joint will be judged by its ability to remain watertight irrespective of the extremes of slew and draw within design limits, or of slight irregularities in the bed, or of any deterioration of the rubber with age. Factors adverse to this requirement can and sometimes do occur in design, manufacture, or installation of the joints. They may be classified as follows:

Design faults

(1) Lack of effective provision for accurately locating the rolling 'O' ring when making the joint.
(2) Lack of means of preventing excessive travel of rolling 'O' rings.
(3) Excessive or inadequate compression or hardness of ring.
(4) Excessive stretching of the ring.

Manufacturing faults

(1) Variation in the width of the annular gap caused by distortion of the socket or spigot or both (ovality) i.e., by excess tolerances on variation in diameter.
(2) Roughness or irregularity of the contact surfaces of the socket or spigot.
(3) Pipe ends not square to the axis.
(4) Contact surfaces of spigot and socket not parallel to each other or at an excessive angle to the pipe axis.
(5) Porosity in the pipes especially in the sockets.

Rubber ring faults

(1) Rubber of inferior quality or inappropriate physical characteristic, e.g., Subject to excessive creep or age hardening.
(2) Variation in the cross section and/or hardness of the rubber rings, either between similar rings or in the same ring.
(3) Rubber cord not properly spliced.

6.5 INSTALLATION FAULTS

(1) Pipe maker's instructions not adhered to or confused with those of other makers, e.g., regarding lubrication.

(2) Rings used after improper storage, e.g., exposed to light and air over a considerable period causing deterioration of the rubber.

(3) Rings used which have been supplied by a different pipe maker or for smaller pipes, e.g., wrong section or hardness, or for excessive stretch.

(4) Rolling rings not placed precisely square to the axis and at the correct distance from the spigot end; rings twisted or stretched unevenly; rings damaged by cuts or abrasions; pipe axis not concentric when jointing; spigots and sockets chipped or otherwise damaged.

(5) Jointing surfaces not properly cleaned; external annular gap not protected against ingress of bedding or other granular material, internal joint gap not cleared of dirt, stones, sand etc.

(6) Unsuitable pulling or pushing tackle used in jointing.

(7) Axial gap between spigot and socket too narrow, or too wide; vertical gaps in any concrete bedding or surround, of inadequate width, misplaced or omitted.

(8) No uniform bedding, allowing one pipe to transmit load to the other and so cause uneven compression in the rubber joint ring, e.g., by disturbance of the bed during jointing.

(9) Socket supported by hard objects, or on the foundation soil.

(10) Inadequate training or experience of the pipe layer.

Table 6.2

Inspection report of mould shell

(For PSC pipe moulds)

Report No:-
Name of party:-
Name of inspector:-

Inspection:-
Order No:-
Date of inspection:-

Report No:-
Name of party:-
Name of inspector:-
Pipe Dia.:-
Date of delivery:-
DRG No and Date:-

THK. _____ mm

Dimensions		C1	D1	D2	D3	D4	D5	D6	D7	C2	Dowel pin			Seam bolt	
											Pin φ	Hole φ	NOS	Hole φ	NOS
Required	Max.														
	Man.														
Tol.	A														
	B														
	C														
	D														

The dimensions which are out of range to be underlined with red colour.
Remarks:-

Sign of Party:-   Rejected/Not accepted/Accepted   Sign of Inspector:-

Table 6.3

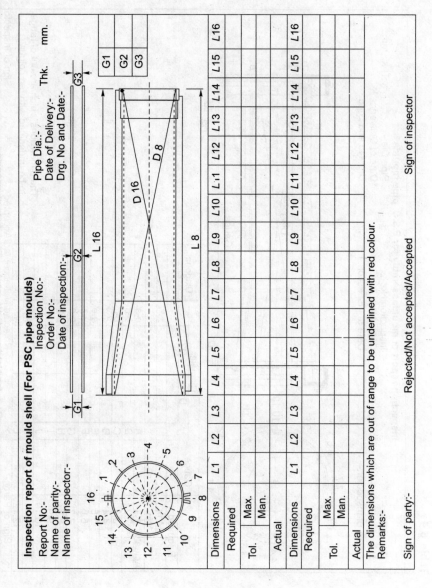

Inspection report of mould shell (For PSC pipe moulds)

Report No:-
Name of parity:-
Name of inspector:-

Pipe Dia.:-
Date of Delivery:-
Drg. No and Date:-

Inspection No:-
Order No:-
Date of inspection:-

Thk.   mm.

Dimensions	L1	L2	L3	L4	L5	L6	L7	L8	L9	L10	L11	L12	L13	L14	L15	L16
Required																
Tol. Max.																
Tol. Man.																
Actual																
Dimensions	L1	L2	L3	L4	L5	L6	L7	L8	L9	L10	L11	L12	L13	L14	L15	L16
Required																
Tol. Max.																
Tol. Man.																
Actual																

The dimensions which are out of range to be underlined with red colour.

Remarks:-

Sign of party:-   Rejected/Not accepted/Accepted   Sign of inspector

Table 6.4

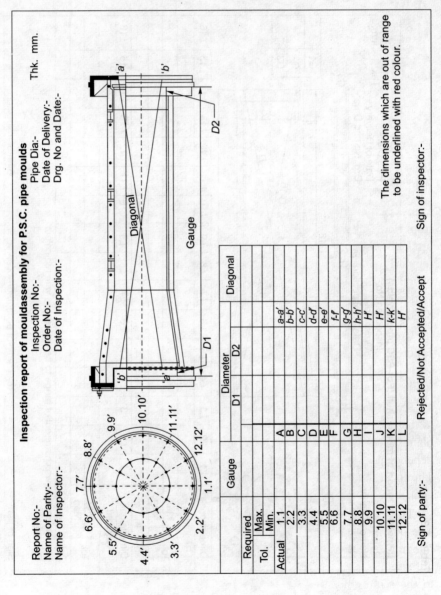

Inspection report of mouldassembly for P.S.C. pipe moulds

Report No:-
Name of Parity:-
Name of Inspector:-

Inspection No:-
Order No:-
Date of Inspection:-

Pipe Dia:-   Thk. mm.
Date of Delivery:-
Drg. No and Date:-

The dimensions which are out of range
to be underlined with red colour.

Gauge	Diameter		Diagonal
	D1	D2	
A			a-a′
B			b-b′
C			c-c′
D			d-d′
E			e-e′
F			f-f′
G			g-g′
H			h-h′
I			H′
J			H′
K			k-k′
L			H′

Gauge			
Required		1.1	
Tol.	Max.	2.2	
	Min.	3.3	
Actual		1.1	
		2.2	
		3.3	
		4.4	
		5.5	
		6.6	
		7.7	
		8.8	
		9.9	
		10.10	
		11.11	
		12.12	

Sign of party:-   Rejected/Not Accepted/Accept   Sign of inspector:-

CHAPTER 7

Requirements of Raw Materials for Pipes

7.1 INTRODUCTION

Many factors have contributed to the success of the concrete pipe industry. Important among these are the ability to use locally available materials, production plants located close to construction projects, and personal services regularly provided by the manufacturer to engineers, contractors and public officials. Include in these services are design and specification assistance, seminars, plant tours and the ability to quickly accommodate the changing deeds of contractors by shipping daily to the project site.

Concrete pipe is manufactured throughout North America. The distance concrete pipe is transported from point of manufacture to the installation site varies. In densely populated area a haul exceeding 75 miles would be unusual, while in less densely populated areas haul distances may exceed 200 miles.

Expansion and improvement of sever systems and highways are closely tied to the needs and economics of an area. Employees and owners of concrete pipe plants usually live near where the pipe is manufactured ad installed. They are dependent on and vitally interested in the services provided by the community. The local nature of the concrete pipe industry allows the pipe manufacturer to contribute to the community and benefit from its expansion.

The quality of precast concrete pipe has also contributed to its success. The quality is obtained from sophisticated facilities, processes and equipment integrated under controlled conditions. While a number of different processes are used, each is capable of producing precast concrete pipe that conform to the requirements of applicable ASTM standards. Discussed briefly in this chapter are the component materials, techniques and equipment used to obtain a consistently high quality product.

The main raw materials for pipes are:

(1) Concrete

(2) Steel

Choice of required quality of raw materials is essential for producing quality pipes.

7.2 CONCRETE

It is a mixture of cement, aggregates, water and may contain one or more admixtures. Basic requirement includes the following:

(i) Cement

(ii) Aggregates

(iii) Mixing

(iv) Compaction

(v) Curing

7.3 IMPORTANT NOTE

Concrete required for pipe is little different from the concrete used for other works. The main difference is the variation of water cement ratio, which is very important for concrete. The initial water cement ratio is about 0.4 to 0.45 and the fine water cement is around 0.3 to 0.32. Judging the final water cement achieved is difficult. Hence, conventional methods of mix design are not totally applicable here. The actual method is of trial and error. Grading of fine aggregate (sand) plays an important role as it has to allow water to come out.

Experience shows that water is relatively easily pushed thorough particles larger than about 600 micron (No 25) in diameter and is substantially held by particles smaller than this, at the same time compact the concrete besides giving a smooth to a comparatively thinner section. It is therefore preferable to have rounded particle in it. Humes of Australia has done considerable work and they have given a curve for grading of fine aggregate.

They have also developed curves for combined aggregates for spinning process.

The main thing noticed is that better compaction is achieved by ensuring that aggregate grading does not have a complete absence of material in two consecution sieves sizes. In fact for example, between 10 mm and 2.36 mm (No 7). In effect any size of particle must act as filter to prevent excessive movement of next smaller size of materials or have continuous grading.

The mix design for vertical casting process is not much different from conventional concrete. But mixing has to be thorough, preferably in a forced one and not a gravity mixer. In both processes the cement is slightly on higher side because fines are more in pipe concrete.

Admixtures are now common in all concrete both chemical and mineral. Mineral admixture contains silicate or aluminates. Strength for a normal concrete is because of calcium silicate or calcium aluminates. In cements which are available, calcium is slightly on the higher side than is required by silicate in it and extra calcium remains as free time. In mineral admixtures which contain silicate or aluminates are added to concrete. These silicates react with the free time and form calcium silicate and increase the strength. But if more quantity of mineral admixture is added they are not so effective. Fly-ash, silica fume Granulated Blast furnace slag (GGBS) are normally used as admixtures. The quantity can be decided only after studying percentage of calcium in the cement silica fume (micro silicate) has more advantage because it falls e surface between aggregates as it is 100 times finer than cement, besides it has higher silica content.

All the above points need special attention, hence those are considered in more details is follow pages:

7.4 (A) PORTLAND CEMENT (OPC)

Certifications of different manufacturing are to Appendix.

7.4.1 Manufacture and Properties

Ordinary Portland cement is a very commonly known and yet a versatile material and is a part of our everyday life. Though it has its limitations due to volume changes, poor tensile strength, and creep, its setting, strength development and case of handling, bond with reinforcement, mould ability and relative imperviousness has made it a favorite construction material.

Cement is scientifically controlled and manufactured product which is far superior to artificial hydraulic lime. The pioneer for cement is John Smenton (1756) who successfully produced a masonry cement for eddy stone light house production. The refinement was made by Aspidin (1825) and he called it "Portland cement", meaning that the hydrated product will be as hard as stone available in Portland district.

Portland cement is defined as a product produced from sintering calcareous and argillaceous constituents and grinding the resulting clinker with permissible amounts of admixtures such as gypsum which modify the properties of the final product to some specific advantage. The main constituents for the manufacture of cement are limestone and clay or shale. Sometimes iron and alumina bearing materials are added as third or fourth component. Cement is made either wet or dry process. The wet process involves grinding to - 170 mesh the raw materials, storing them as slurry, proportioning them and sintering in rotary kilns using coal or fuel oil. The dry process involves initially drying the materials to remove quarry moisture, grinding to 170 mesh, storing dry, proportioning and sintering as in wet process. The advantages and disadvantages of the two processes are considered beyond the scope of this.

The sintering or clinkering takes place at a temperature around 2400°F (1400°C) to enable formation of the desired compounds in the clinker. Rotary kilns are used which vary in length from 30 to 110 meters in 'dry process' and extend up to 200 meters in wet process. The diameter of kiln varies from 2 to 4 meters and these rotary kilns rotate at a speed of 30 to 100 revolutions per hour in an inclined pitch of 20 mm to 80 mm per meter. In the kiln the initial processes involve dissociation of $MgCO_3$ at 330°C, of $CaCO_3$ at 900°C and combination of iron, alumina and silica with lime at temperatures over 1000°C up to 1400°C. The clinker that is formed in the kiln is quickly cooled, stored and aerated, ground with gypsum, and bagged in water-proof bags or jute bags.

Cement contains four major compounds:

(1) C_3A -   Tricalcium aluminate
(2) C_3S -   Tricalcium silicate
(3) C_2S -   Dicalcium silicate
(4) C_4AF -   Tetracalcium aluminoferrite

Indian cements have a range of these compounds as given below:

(1) C_3A -   7 - 12%
(2) C_3S -   35 - 55%
(3) C_2S -   20 - 40%
(4) C_4AI -   5 - 10%

The individual constituents are generally in the following ranges:

CaO   -   62 - 65%
SiO_2 -   19 - 23%
Al_2O_3 -   4 - 7%
Fe_2O_3 -   2 - 4%
SO_3 -   1 - 2.5%
MgO   -   1 - 4%
Free Lime   -   0.5 - 1%

7.4.2 The Role of Cement Compounds

During the formation of compounds in the hot zone of kiln the order of formation is O_4AF, C_3A, C_2S and C_3S.

C_3A liberates lot of heat (2 cal/gm) and contributes to the setting and early strength development. C_3S contributes heat (1.1 cal/gm) as well as a large part of strength up to 28 days. C_2S contributes less heat (0.5 cal/gm) but is responsible for strength mostly after 28 days. It is also responsible for the autogenous healing of concrete. C_4AF has no special significance but this compound and C_3A help as fluxes in the formation of C_3S. MgO if présent as Periclase can cause slow expansion. Periclase is formed by slow cooling of clinkers. Free time can also cause unsoundness if present over 4% free lime results from incomplete reaction in the kiln. Under-liming may result from improper mix control and is undesirable since it produces cement which will have low strength development properties. It may be noted that each 1% increase in $CaCO_3$ content in ran mix increases C_3S by 14.5% and decreases C_2S by 13.5%. Rapid hardening cements have higher C_3S and low heat cements have higher C_2S.

7.4.3 Properties

Specific gravity of cement is generally closer to 3.1. The specific gravity value is utilised in design of concrete mixes. It may be noted that this specific gravity value changes in cement paste, which has a lesser volume than the individual volumes of cements and mortar.

7.4.3.1 Fineness

The clinker along with gypsum is ground in mill, to ensure that the cement passes the requirement of I.S.269 (Retention on 170 mesh 10% and surface area 2250 sq.cm/gm. Blaine). Indian cements have a fineness of about 2600 – 3500 sq. cm/gm.

Increased fineness of cement increases the strength particularly at early age, decreases bleeding, decreases autoclave expansion and improve the workability and cohesiveness of concrete. The sieve method gives only an idea of the oversize material and does not indicate, the particle size distribution of the - 170 fraction. In the air permeability method (Blaine) the permeability of a cement bed to the flow of a known quantity of air through the compacted bed of cement of predetermined porosity is measured. Particle size distribution in determined by turbidity meter technique (Magnor) and is more useful to differentiate cements of equal specific surface having different bleeding rates or strength value. About 60% of surface area of Portland cement is attributed to articles below 7.5 microns. The percentage passing this size may be about 25% and this fraction contributes to the desirable properties. Coarser frictions range up to 45 microns. Particles larger than 45 microns have no significance and may extend up to 75 microns. About 85% of cement passes 325 mesh or 60 microns sieve. An increase in fineness from 1800 to 2500 cm^2/gm (Magnor) increases one day strength of concrete by 50 to 100%, at 3 days from 30 to 60% and at 7 days by 15 to 40%. Increased fineness increases coverage of paste and thereby the efficiency. Very fine grinding increases the water requirements and gypsum requirements. C_3S is most in finer sizes and C_2S in coarser sizes. The C_3A and C_4AF are distributed in all frictions.

7.4.3.2 Setting

The setting time of cement and concrete in controlled by the crystalline C_3A which causes flash set if it is not retarded by gypsum. If the retardation is done, the setting time is controlled more by the C_3S. C_2S has no effect on setting time. The nature of setting and hardening of cement paste is explained by a combination of gel theory and colloidal theory.

The setting of cement is determined on a plate worked up to a uniform consistency. The consistency is determined by the penetration of a weighted 1 cm diameter plunger to a known depth. (Vic at apparatus). The initial and final sets are denoted by the extent of penetration of a 1 mm × 1 mm square needle and are checked at intervals. Normally our cements have an initial sets of over 90 minutes and final not of over three hours. Since the setting time of neat paste has no direct correlation with setting of mortars or concrete and serves merely as a general index.

7.4.3.3 False set

It is essentially due to the hot-grinding of clinker or rise in melt temperature. The stiffening is caused due to the gypsum present converting to plaster of parries during grinding which hydrates to gypsum when water is added. Sometimes flash set is attributed to the carbonation of the alkali oxides present in cement. False set can cause increased water requirements, reduced strength, reduced bond etc. in concrete. False set does not evolve heat and can be overcome by additional mixing. With flash set heat is liberated by C_3A and the plasticity is not recoverable. Flash set is due to un-retarded C_3A.

7.4.3.4 Soundness

Unsoundness in cement is essentially due to excessive free lime and MgO. This is detected by measuring expansion of cement-pastes, gauged for a specific consistency, aged for 24 hours and steam cured. The simpler test is the Le-Chatelier ring expansion test in which an explanation of 10 mm is considered as indicative of unsoundness.

In the autoclave expansion test bars of size 1" × 1" × 10" are subjected to steam pressure at 295 psi which accelerates the hydration of dead burnt uncombined lime or crystalline magnesia. An expansion of +.0.5% indicates unsoundness.

7.4.3.5 *Compressive and tensile strength*

The I.S. 269 stipulates these tests as a minimum of the quality of cement produced. Fresh cement is expected to yield the following strengths as minimum in 1:3 motors.

	Compressive strength	Tensile strength (optional)
3 - days	115 kg/cm^2	20 kg/cm^2
7 - days	175 kg/cm^2	25 kg/cm^2

The tests are conducted under standard conditions using standard sand (I.S.650) producing a mortar of specified consistency, vibrated or tamped to form cubes or briquettes. The method given is intended to achieve uniform strength rather than absolute strength development and as such can not be considered to be related to performance in field concrete or mortar.

Storage of Cement

Even well stored cement can lose strength by 20, 30, 40 and 50% when stored for 3, 6, 12 and 24 months. Good storage involves a weather proof shed, dump proof for floor and closes staking to prevent circulation of air; stacking must be in the order of receipt and issue also must be in the same order as received. Proper storage and issue includes, clear labeling of consignments, number of bags in consignment, date of receipt at rail/road head and chronological identity for facilitating issue. Stack height should not be more than 10 bags. This will prevent formation of hard air-set lumps.

Weight of cement bags

Normally it is presumed that one bag in equivalent to 1.25 cft. This conception is likely to be erroneous since cement can vary in its bulk density from 70 to 100 pounds per cubic foot depending upon its compact nose during filling. The weight of cement in a bag, which is 50 kg as bagged can also vary considerably due to methods of handling and transporting. The effect of such fluctuation as a factor influencing variation in quality of concrete will be discussed later.

Sampling of cement

The details of sampling cement from stocks or from running consignments are covered in I.S. 269. It may, however, be noted that one sample of five pounds (2.5 kg) shall be taken from 12 separate bags from each lot of 1000 bags. The samples are taken with the use of a sugar sampler and about 200 gm is collected from each bag and mixed up.

Variability in cement quality

Some of the variation in the strength of concrete is attributable to variation in the quality of cement. Some feel that if the cement passes the minimum standard of requirement of a specification it should be good enough to produce concrete of uniform quality and strength. Small variations in quality in the day to day production is inevitable since variation in the argillaceous or calcareous raw material and fuel are inevitable in spite of extremely careful proportioning in the factories.

Wright states that "the variation between cement from different works will enhance the variation of the works cube strengths and therefore obtaining cement from more than one factory should be discouraged". Wing and Rutgers found that 65 to 80% of the total variation of 28-day strengths was attributable to variation in specific surface and tricalcium silicate content of cement. Sparkers opines that cement from only one mill should be used to obtain more uniform product.

7.5   PORTLAND POZZOLANA CEMENT

This is the name give to interground or blended mixture of Portland cement clinker and granulated blast furnace slag. The proportion of slag is limited to 65% by I.S. 455 - 1962.

PBFS Cements are similar to OPC. Even specification requirements regarding physical properties are the same. As a matter of fact the fineness of PBFS cement is higher and the heat of hydration is lower than OPC.

7.6   RAPID HARDENING PORTLAND CEMENT

This is characterized by its potential to develop high early strengths. The C_3S content and the fineness in normally higher than OPC. The setting and soundness limits are the same as OPC.

7.7   COMPARATIVE STATEMENT

Remarks

Figures in brackets are based on the revision in 1967 of the relevant specifications. For compressive strength tests graded standard sand 2.0 mm to 0.09 mm is specified as per revision instead of the 0.85-0.60 mm grade. The water content for preparing the cubes in also revised. The tensile strength test is omitted and the transverse strength on prisms is included as a now optional test. As regards chemical composition, the Mgo content is permitted to be up to 6% as against 5% in OPC.

Table 7.1 The following statement gives comparatives important physical requirements of these cements

I.S. Specification	269-1958	269-1958	455-1962	1489 1962
Type of Cement	OPC		PBFS	PPZ
Fineness:				
(a) Residue on 90 micron I.S. sieve, max.%	10	5	10	5
(b) Specific surface minimum (Blaine) cm^2/g		3250	2250	3000
Setting time				
Initial (Min.)	30	30	30	30
Final (Min.)	600	600	600	600
Compressive strength (kg/cm^2)_____				
1day	-	115(160)	-	-
3 days	115(160)	210(275)	115(160)	-
7 days	175(220)	-	175(220)	140 (175)
14 days	-	-	-	210(150)
Tensile strength (kg/cm^2) (Optional)				
1 day	-	20	-	20
3 days	20	30	20	-
7 days	25	-	25	-
Soundness				
(a) Expansion by Le-Chatelier method (mm)	10		10	
(b) Expansion by Auto-method (mm)	0.5(0.8)	0.5(0.8)	0.5(0.8) clave	

Certificates of Cement and Fly ash are enclosed.

Test certificates of different cements and flyash are enclosed for comparison.

7.8 ADMIXTURES

Any material deliberately added to concrete before or during mixing other than cement, water and aggregates is called an admixture. There are many types of admixtures; the most common types are:

Air – entraining agents
Accelerators
Retarders
Water reducers
High range water reducers (superplasticizers)
pozzolans

Air-entraining agents

These admixtures produce tiny air bubbles in concrete. The bubbles are formed by mixing action and the air-entraining agents keep the bubbles from breaking up. Properly air-entraining concrete greatly improves concrete's ability to withstand freezing and thawing. Air-entrainment also makes concrete and helps to reduce bleeding and segregation.

Accelerating admixtures

Accelerators speed up the setting and hardening of concrete. They are especially useful in cold weather because concrete hardens very slowly at temperatures below, about 10°C. the most common of these admixtures is calcium chloride. Too much calcium ehloridefe[5] enerete can cause corrosion of reinforcing steel.

Retarders

Retarding admixtures slow down the initial setting of concrete. They are often used in warm weather to keep the concrete from setting before it can be placed and finished. Most retarders are also water reducers. They do not reduce slump loss.

Water reducers

As the name suggests, theses materials reduce the amount of water needed to produce a cubic meter of concrete of a given slump. If the amount of water is not reduced, the water reducer will act to increase the slump of concrete. As noted under "retarders", most retarders are also water reducers, so to offset the retarding action, some water reducers contain accelerators.

High range water reducers

These materials are commonly called "superplasticizers" or simply referred to as "supers". They reduce the water requirement for concrete dramatically or they can be used to increase the slump of very stiff concrete. However, the action of these materials lasts only about 30 minutes at normal temperatures-then the concretes stiffen very rapidly.

7.8.1 Pozzoloans

These materials are named for a town in Italy where ash from Mt. Vesuvius was first mixed with lime, water and stone by the Romans to make a form of concrete. Pozzoloans by themselves have no cementing ability but when missed with water and calcium hydroxide (lime) or Portland cement has some cementing ability. Today the most commonly used Pozzoloans are fly ash, slag and silica fume. Pozzoloans react slowly so when concrete contains pozzoloans, longer curing periods are recommended.

7.9 AGGREGATES

Sand, gravel, crushed stone or similar materials, which are mixed with cement and water to make concrete, are called aggregates. These materials make up 60 to 80% of volume of concrete. A cubic meter of normal weight concrete may contain 1500 kgs to 1900 kgs of aggregates.

A number of laboratory tests can be made to find out if an aggregate can be used to make strong durable concrete. Assuming that an aggregate can be used to make good concrete, the important factors that affect how the concrete acts are:

(1) Maximum size

(2) Gradation

(3) Particle shape

(4) Organic impurities

(5) Silt and clay content

Amount of coarse and fine aggregate in the mix

Maximum size: If all of the particles are smaller than about 4.75 mm the aggregate is called fine aggregate. Most fine aggregates are natural sands but some are produced by crushing rock. If most of the particles are larger than about 4.75 mm the aggregate is called coarse aggregate.

Concrete that is made without Ally coarse aggregate is usually called mortar. Most of the concrete that is used in pipe industries has a maximum aggregate size of 20 mm or 25 mm. In heavy construction work and in pavements, larger aggregate sizes are often used, and aggregates as large as 150 mm have been used in concrete for large dams.

The amount of cement and water needed in a mix depends on the aggregate size.

For example, with 25-mm maximum aggregate, less cement and water are needed to produce concrete with a given strength than with 12-mm size aggregate. However, the largest particle size should not be more than:

(a) One-fifth the dimension of non-reinforced members such as walls, steps or footings.
(b) Three-fourths of the clear spacing between rebars or between rebars and forms.
(c) One-third the depth of slabs.

Gradation: Aggregates are made up of particles of many different sizes. To make concrete batches that are essentially the same, the aggregate amount and particle sizes must be essentially the same in one batch as in the next. To measure the particle sizes, a dry sample of the aggregate is passed through a number of sieves starting with the largest openings and using smaller openings in successive sieves. The percentage of the weight passing each sieve is called the gradation.

Standard sieves have square openings. Those that are 4.75 mm or larger are given in terms of the opening size such as 20 mm or 10 mm. Sieves with openings smaller than 4.75 mm are given in terms of mm.

Following is a typical gradation of sand suitable for concrete:

Table 7.2

IS Sieve Size	% Passing	% Each
4.75	100	8
2.75	92	24
1.16	68	25
600 μ	43	22
304 μ	21	15
150 μ	6	6

Particle shape: Ideally, aggregate particles should be shaped like cubes, but without perfectly smooth surfaces. If particles are flat or like slivers, the concrete will tend to be harsh and may be difficult to finish.

Organic impurities: If a sand or gravel contains organic impurities, it may change the setting time of concrete or the concrete may be weaker. There is a simple laboratory test that can be used to check if sand is contaminated with organic material.

Silt and clay content: Materials such as silt and clay will pass through a No 200 sieve. If more than 5% of a sand sample passes through a No 200 sieve, more water may be needed to make a cu meter of concrete, and the concrete may be weaker and the surface less wear resistant. Fine aggregates should have no more than 3% clay lumps.

7.10   MIXING

Thorough mixing is essential for the production of concrete of uniform quality. Mixing must be such that the various sizes of aggregates are uniformly distributed throughout the mix, each aggregate particle being coated with a cement paste of uniform consistency. Mixing may be done by hand or machine.

Machine mixing

There are five main types of concrete mixers commonly in use. While their methods of operation differ considerably, the following general principles are usually applicable

Cement, sand and coarse aggregate should ideally be fed into the mixer simultaneously and in such a manner that the flow of each extends over the same period. The concrete produced in this way is more uniform than that obtained when the ingredients are introduced one after the other.

With most mixers it will be found that filling the loading hopper in more or less horizontal layers of stone, cement and sand, possibly with another layer of different size of stone on top, provides the best results. Placing of the stone at the bottom of the loading hopper is more likely to result in a self-cleaning process, so avoiding the formation of a hardened layer of material at the bottom of the hopper. The piling-up of any one size of material in the throat of the loading skip should be avoided. With large mixers (3 m^3 or larger), the loading sequence of materials into the mixer is of particular importance.

Head packing may occur when the finer materials are fed into the mixer first and become lodged in the head of the mixer. If water or cement is fed in too fast or too hot, cement balling may occur.

The water should enter the mixer at the same tinie and over the same period as the other materials.

When this is not possible, it is advisable to start the flow of water a little in advance of the other ingredients.

If all the water is added before or after the other ingredients, the wetness or slump of the concrete is liable to vary within the batch. The direction of inflow of water may also influence the effectiveness of mixing.

Mixing should continue until the concrete is of uniform consistence, colour and texture. With drum mixers it is an advantage if the mixer driver can see into the drum so that he can observe the consistence and make any necessary minor corrections to the water added.

The mixer should not be loaded beyond its rated capacity. Overloading results in spillage of materials and slower or incomplete mixing, in addition to imposing undue strain on mechanical parts.

The mixer should be set up accurately so that the axis of rotation of the mixer drum is horizontal except in the case of the tilting-drum type. Inaccurate setting may result in poor mixing. Horizontal water measuring tanks and some mass measuring systems are also affected by errors in leveling.

The mixer should be operated at the correct speed as stated by the manufacturer. The speed should be checked regularly.

Some mortar from the first batch of concrete mixed is left behind on the blades and round the drum after the concrete has been discharged. To ensure that this first batch is not too stony, the stone content should be reduced slightly or the mixer fined with a mortar of similar proportions to that of the concrete.

Hardened concrete adhering to the blades and the inner surface of the drum reduces mixing efficiency. Regular cleaning at the end of each shift or where delays in mixing in excess of about 45 minutes are expected is necessary to prevent concrete build-up, especially if stiff mixes are produced. The mixer can be cleaned by loading it with a quantity of stone and water, running it until the drum and blades are clean and then emptying.

Prolonged mixing may lead to the following:

Some mix water evaporates from the mix causing a decrease in slump. If the aggregate is soft, grinding occurs which produces a finer material and reduces workability. Concrete temperature increases due to friction. Some entrained air is lost from air-entrained concrete.

7.10.1 Mixer Types

Mixer types commonly used in south Africa include non-tilting drum, tilting drum, reversing drum, split drum and pan. Through or continuous mixers are rarely used and are not consider in this section.

7.10.1.1 Non-tilting drum

This type has a single drum that rotates about a horizontal axis. The drum is mounted on rollers and driven via a rack and pinion drive or chain. The mixer is charged from one end and mixing is effected by cup – type blades that lift and drop the material as the drum revolves. The drum is discharged from the opposite end by inserting a chute that deflects the concrete out of the mixer as the concrete fall from the mixing blades.

Drum capacity typically ranges from 300 to 1200.

7.10.1.2 Tilting drum

Small tilting drum mixers are commonly used because of their uncomplicated construction and ease of maintenance. Capacities typically do not exceed 200 and gauge boxes are usually required fro batching. Some larger models are fitted with hoppers for charging the mixer.

Mixing is by lifting and dropping the material, some assistance being given by the drum shape that moves the mass along the axis of the drum. Tilting the drum discharges the concrete rapidly.

7.10.1.3 Reversing drum

This type rotates in one direction for mixing and in the opposite direction to discharge the concrete. Typical capacity is 400 to 600e. the drum has to sets of fixed blades, one for mixing and other the other for discharging. When the drum is reversed after mixing, the discharge is quick and complete. Mixing times are shorter than for tilting and non-tilting mixers. Very little built-up occur on the blades.

Reversing drum mixers are usually fitted with integral loading mass – measuring hopers.

A truncated cone is generally fitted to the discharged end of the mixer to channel the flow of the concrete into wheelbarrows, skips, dumpers, etc.

Badly worn and bent blades reduce effectiveness of mixing and should be replaced. Adjustable mixer and scraper blades must be kept set to the correct clearances. Wear of the inlet and discharge chutes eventually results in loss of materials. These chutes should therefore be kept in good repair.

Rubbing grease or oil over the mixer after cleaning prevents cement building up on the outside of the mixer. Layers of cement are particularly liable to build up in the nose of the loading hopper and should be chipped off regularly.

7.10.1.4 Pan

Pan mixers are available for site use in sizes typically ranging from 200 to about 2000 literes. these forced action mixers thoroughly mix even low-slump and lean concrete quickly with mixing times often well under a minute. Because of this they are ideal for use in precast works.

Increasing use is being made of these mixers in the ready mix industry. The capital cost of such mixers is high, careful, and frequent maintenance is needed to ensure efficient operation.

There are two types of pan mixers. One has a circular rotating pan with mixing spiders eccentrically to the pan. The spider may revolve in the same direction as, or counter to, the pan. The discharge doors are located centrally in the floor. The other has a stationary circular pan with central mixing arms extending from a central gearbox. The mixing blades plough the materials at high speed. The discharge doors are located on the outer edge of the pan.

7.11 UNIFORMITY OF MIXING

The proportions of constituent materials can vary considerably within a batch., but these variations are difficult to measure. Such variations may be due, in rotating drum mixer to the lack of movement of material from one end of the drum to the other. ASTM C944-C.94M requires samples to be taken from the first sixth and fifth sixth of the discharge and that the differences in the properties of the two samples should not exceed any of the following:

Density of concrete: 16 kg/m^3

Air content: 1% of the volume of concrete

Slump: 25 mm when average is less than 100 mm; 38 min when average is 100 to 150 mm

Aggregate retained on 4,75-mm sieve: 6%

Density of air-free mortar: 1,6%

Compressive strength (average of three cylinders at seven days): 7,5%

7.11.1 Mixing time

In order to use the mixer to maximum efficiency it is necessary to know the minimum mixing time necessary to produce a concrete of uniform composition.

The mixing cycle includes:

The time taken to charge the mixer

Actual mixing time (measured from when all the ingredients have been added to the mixer)

Discharge time: Optimum mixing time depends on the type and size of mixer, speed of rotation and method of blending ingredients during charging of the mixer. While mixing and discharge are taking place, the hoppers should be charged with material for the following batch to save time.

The mixing cycle should never be reduced in order to increase production. Reducing the cycle shortens the actual mixing time and this may result in concrete with highly variable workability and strength properties. The time of mixing should not be less than that recommended by the manufacturer of the mixer or than that otherwise determined as being satisfactory.

7.12 TEST CERTIFICATES OF CEMENT, FLY ASH, BLAST FURNACE SLAG CEMENT ARE IN FOLLOWING PAGES

See Appendixes A and B.

Appendix - A

Properties of Cement		
Particulars	Test	Requirements
	Results	IS : 12269-1987
Chemical Requirements		
CaO - 0.7 SO_3	0.908	0.8 to 1.02
$2.8SiO_2 + 1.2\ Al_2O_2 + 0.65\ Fe_2O_3$		
%Al_2O_3 / % Fe_2O_3	1.06	0.66 Min.
Insoluble residue (% by mass)	1.49	3.0 Max.
Magnesia (% by mass)	1.06	6.0 Max.
Sulphuric Anhydride	1.51	3.0 Max.
Total loss on ignition(%)	1.92	4.0 Max.
Chloride (%)	0.028	0.1 Max.
Physical Requirements		
Fineness (m^2 /kg)	297	225 Min.
Normal consistency	28.25%	
Setting time Initial (min.)	190	30 Min.
Setting time Final (min.)	275	600 Max.
Soundness-Le Chatelier Expansion	0.2	10 Max.
Soundness-Autoclave Expansion	0.01	0.8 Max.
Compressive strength		
3 days	42	27 Min.
7 days	51.6	37 Min.
28 days	61.2	53 Min.
Properties of Fly Ash		
Description	Results	Requirements of
		IS 3812 :2003
Chemical Properties		
Silicon Dioxide, SiO_2(%)	59.98	35
$SiO_2 + Fe_2O_3 + Al_2O_3$ (%)	94.11	70
Magnesium Oxide (%)	0.48	5
Sulfur trioxide (%)	0.04	5
Loss of Ignition (%)	0.82	5
Total alkali, Na_2O Eqv. (%)	0.89	1.5
Physical Properties		
Retained on 45 μ Sieve (%)	3.64	50
Autoclave Expansion, (%)	0.05	0.8

Appendix - B

JSW CEMENT LTD

TEST CERTIFICATE

Graund Granulated Blast Furnace Slag

IS No	Characteristics	Requirement as Per BS: 6699	Test Result
1	Fineness (M^2/kg)	275 (Min)	407
2	Insoluble Residue (%)	1.5 (Max)	0.57
3	Magnesia Content (%)	14 (Max)	9.00
4	Sulphide Sulphur (%)	2 (Max)	0.47
5	Sulphide Content (%)	2.5 (Max)	0.10
6	Loss on Ignition (%)	3 (Max)	0.20
7	Manganese Content (%)	2 (Max)	0.05
8	Chloride Content (%)	0.1 (Max)	0.01
9	Glass Content (%)	67 (Min)	88
10	Moisture Content (%)	1 (Max)	0.12
11	Chemical Moduli		
A	$CaO + MgO + SiO_2$	66.66 (Min)	74.9
B	$CaO + MgO / SiO_2$	>1.0	1.32
C	CaO / SiO_2	<1.40	1.04

Note: There is Indian Standard on GGBS. For reference BS specification is given. Granulated Slag used for Steel Ltd., GBS conform to IS 12089 : 1987

Week No: 37

07-0.9-2009-13-09-2009

Authorised Signatory

ACW-'UltrTech 53'	Issue Date: 31.05.2013

ADITYA BIRLA

UltraTech

UltraTech Cement
TEST CERTIFICATE
FOR
53 Grade Ordinary Portland Cement
(Fly Ash based)
Test & Compliance

IS 12269

CM/L-2058041

Particulars CHEMICAL REQUIREMENTS	Test Results	Requirements of IS: 12269-1987	
1. $\dfrac{CaO-0.7SO_3}{2.8\ SiO_2 + 1.2Al_2O_3 + 0.65\ Fe_2O_3}$	0.92	0.80 1.02	Min Max
2. Al_2O_3/Fe_2O_3	1.32	0.66	Min
3. Insolube Residue (% by mass)	1.19	3.00	Max
4. Magnesia (% by mass)	2.49	6.00	Max
5. Sulphuric Anhydride (% by mass)	1.53	3.00	Max
6. Total Loss on Ignition (% by mass)	1.67	4.00	Max
7. Total Chlorides (% by mass)	0.010	0.10	Max

PHYSICAL REQUIREMENTS

1. Fineness (m^2/kg)	315	225	Min
2. Standard Consistency (%)	28.0		
3. Setting Time (minutes)			
a. Initial	155	30	Min
b. Final	240	600	Max
4. Soundness			
a. Le-Chat Expansion (mm)	1.0	10.0	Max
b. Autoclave Expansion (%)	0.088	0.8	Max
5. Compressive Strength (MPa)			
a. 72 +/– 1 hr. (3 days)	42.0	27	Min
b. 168 +/– 2 hr. (7 days)	51.5	37	Min
c. 672 +/–4 hr. (28 days)	64.0	53	Min
6. Performance Improver (%)		5.0	Max
a. Limestone	2.0		
b. Fly Ash	NA		
c. Granulated Slag	NA		

The above cement complies with the requirements of
IS: 12269-1987 for 53 Grade Ordinate Portland Cement

Date of Dispatch: 16.04.2013
Week no: 16 of 2013

HOD(QC)

UltraTech Cement Limited
Works: Awarpur Cement Works, Awarpur, Taluka-Korpana, District-Chandrapur, Maharashtra-442 917

ACW-'UltrTech PPC'

Issue Date: 31.05.2013

ADITYA BIRLA

UltraTech

UltraTech Cement
TEST CERTIFICATE
FOR
Portland Pozzolana Cement
(Fly Ash based)
Test & Compliance

IS 1489

(Part)
CM/L-1280946

Particulars CHEMICAL REQUIREMENTS	Test Results	Requirements of IS: 1489-1991(Part 1)	
1. Insolube Material (% by mass)	22.39	$X + \dfrac{4.0(100-X)}{100}$ (x = Declared % of Fly Ash)	Max
2. Magnesia (% by mass)	1.98	6.00	Max
3. Sulphuric Anhydride (% by mass)	1.63	3.00	Max
4. Loss on Ignition (% by mass)	1.06	5.00	Max
5. Total Chlorides (% by mass)	0.011	0.10	Max
PHYSICAL REQUIREMENTS			
1. Fineness (m^2/kg)	341	300	Min
2. Standard Consistency (%)	31.5		
3. Setting Time (minutes)			
a. Initial	250	30	Min
b. Final	335	600	Max
4. Soundness			
a. Le-Chat Expansion (mm)	1.0	10.0	Max
b. Autoclave Expansion (%)	0.040	0.8	Max
5. Compressive Strength (MPa)			
a. 72 +/– 1 hr. (3 days)	30.0	16	Min
b. 168 +/– 2 hr. (7 days)	40.0	22	Min
c. 672 +/– 2 hr. (28 days)	54.0	33	Min
6. Drying Shrinkage (%)	UT	0.15	Max
7. % of Fly Ash Addition	27.0	15.0	Min
		35.0	Max

The above cement complies with the requirements of
IS: 1489-1991 (Part 1) for Portland Pozzolana Cement (Fly Ash based)

*Amount of Fly Ash addition in finished cement does not vary
more than declared value.

Date of Dispatch: 16.04.2013
Week no: 16 of 2013

HOD(QC)

UltraTech Cement Limited
Works: Awarpur Cement Works, Awarpur, Taluka-Korpana, District-Chandrapur, Maharashtra-442 917

ADITYA BIRLA

UltraTech

UltraTech Cement
TEST CERTIFICATE
FOR
43 Grade Ordinary Portland Cement
Test & Compliance

IS 8112

CM/L-2058142

Particulars CHEMICAL REQUIREMENTS	Test Results	Requirements of IS: 12269-1987	
1. $\dfrac{CaO-0.7SO_3}{2.8\,SiO_2 + 1.2Al_2O_3 + 0.65\,Fe_2O_3}$	0.92	0.66 1.02	Min Max
2. Al_2O_3/Fe_2O_3	1.34	0.66	Min
3. Insolube Residue (% by mass)	1.46	3.00	Max
4. Magnesia (% by mass)	2.52	6.00	Max
5. Sulphuric Anhydride (% by mass)	1.46	3.00	Max
6. Total Loss on Ignition (% by mass)	1.95	5.00	Max
7. Total Chlorides (% by mass)	0.011	0.10	Max
PHYSICAL REQUIREMENTS			
1. Fineness (m^2/kg)	281	225	Min
2. Standard Consistency (%)	27.5		
3. Setting Time (minutes)			
a. Initial	175	30	Min
b. Final	250	600	Max
4. Soundness			
a. Le-Chat Expansion (mm)	1.0	10.0	Max
b. Autoclave Expansion (%)	0.092	0.8	Max
5. Compressive Strength (MPa)			
a. 72 +/– 1 hr. (3 days)	35.5	23	Min
b. 168 +/– 2 hr. (7 days)	45.0	33	Min
c. 672 +/– 2 hr. (28 days)	55.0	43	Min
6. Performance Improver (%)		5.0	Max
a. Limestone	3.0		
b. Fly Ash	NA		
c. Granulated Slag	NA		

The above cement complies with the requirements of
IS: 8112-1989 for 43 Grade Oridinary Portland Cement.

Date of Dispatch: 16.04.2013
Week no: 16 of 2013

HOD(QC)

UltraTech Cement Limited
Works: Awarpur Cement Works, Awarpur, Taluka-Korpana, District-Chandrapur, Maharashtra-442 917

Fly Ash in India

- Indian subcontinent possesses billions of tons of sub-bituminous coal reserves

- In modern thermal plants operations, coal is segregated of its impurities and clayish lumps and ground to a fineness of about 7,5 mm (70% passed)

- Almost all boilers generae Class F fly ash, except in a couple of pockets in South India and extreme West where Class C fly ash is generated.

Quality of Fly Ash in India-Dahanu Fly Ash

Description	Results	Requiremens of IS 3812:2003
Chemical Properties		
Silicon Dioxide, SiO_2 (%)	59.98	35
$SiO_2 + Fe_2O_3 + Al_2O_3$ (%)	94.11	70
Magnesium Oxide (%)	0.48	5
Sulfur trioxide (%)	0.04	5
Loss of Ignition (%)	0.82	5
Total alkali, Na_2O Eqv. (%)	0.89	1.5
Physical Properties. (%)		
Retained on 45 μ Sieve (%)	3.64	50
Autoclave Expansion, (%)	0.05	0.8

CHAPTER 8

Loads on Buried Pipes

8.1 CHARACTER OF EXTERNAL LOADS ON BURIED PIPES

External loads ordinarily control the design of a sewer pipe ring. Buried pipes may be subjected to such forces as are indicated in Figs. 8.1 and 8.2. In addition to the forces and conditions affecting the loads on buried pipes as indicated in the figures, there may be included such factors as the length of the trench, the flexibility of the pipe material and the distribution of the load.

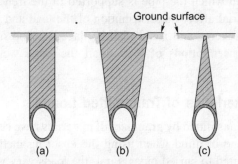

Fig. 8.1 Hypothetical load transmission to buried pipes

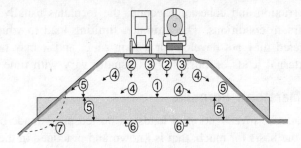

1. Dead Load
2. Static Super-load
3. Impact
7. Undermining
4. Transverse Earth Pressure
5. Frost
6. Soft, Uneven Foundation

Fig. 8.2 Loads on buried pipes

If the external loading is uniform and concentric around the circumference of a "circular" pipe, the internal stresses in the wall of the pipe will all be compressive. If the loading is not uniform or the pipe is not uniform or the pipe is not circular, all possible forms of internal stress may exist.

The flexibility of the pipe affects the loads and stresses, because the change in the shape affects the distribution and intensity of external loads. The load on a rigid pipe may be somewhat as indicated in Fig. 8.1, whereas the deflection of the top of a flexible pipe might cause arching in the surrounding ground with a load distribution somewhat as indicated in Fig. 8.2. Marston has defined a "rigid" conduit in which a change of more than 0.1% of its horizontal or its vertical dimensions would cause definitely injurious cracks; for semi-rigid conduits the limits are set at more than 0.1% but less than 3%; and for flexible conduits the limits are set at more than 3%. Vitrified-clay, concrete and cast-iron pipes are rigid pipes. Steel pipes are flexible.

The nature of the load placed on the ground surface over the pipe, called a superimposed load, affects the intensity of load transmitted to the pipe. Such loads may be classified as concentrated, in truck wheel loads for instance, or as distributed, as in piles of construction materials on the surface.

The manner in which the pipe is supported in the trench, and the nature of the backfill material affect the distribution of the load and the internal stresses. Boulders, broken stone, and rough debris placed on top of a buried pipe will cause greater concentrations of load, than the same weight of dry sand or wet clay.

8.1.1 Characteristics of Transmitted Loads

Loads due to or transmitted by granular fill materials have certain characteristics, that must be borne in mind when using the ensuring methods in determining the loads transmitted to buried pipes. First, the loads vary with variations in the properties of the materials, such as weight, settlement, moisture, temperature, internal friction and cohesion; second, the formulas usually applied represent only ultimate conditions. The ultimate, limiting load to which a conduit may be subjected may not develop for a long time, and it may never develop; and third, external loads on buried pipes usually vary with time.

8.1.2 Marston's Formula

The exhaustive investigation of loads on buried pipes made by Anson Marston remains the basis for much, that is known and practiced in the field. Conditions under which conduits are buried may be classified as:

(1) completely buried in a ditch, with undisturbed ground as the bottom and side walls of the ditch - this is called a "ditch condition"; or (2) partly buried in a shallow ditch, the remaining part of the conduit being covered with an

earth embankment or backfill projecting above the surface of the surrounding ground. This is called a "projection condition" or projection conduit". If there is a load on top of the backfill it is called a "superimposed load". A moving vehicle or a concentrated load which does not extend along the ditch is called a "short load". A load extending along the ditch, such as a pile of building material, is called a "long load".

8.2 LOAD IN TRENCH CONDUITS

When the conduit is placed in a trench not wider than two or three times its outside breadth, and covered with earth, the backfill will tend to settle downward. This downward movement or tendency for movement of the backfill in the trench above the conduit, is retarded by frictional forces along the side of the trench which act upward on the backfill in the trench and help support the backfill.

The vertical frictional forces are equal to the active lateral pressure by the earth backfill against the side of the trench multiplied by the tangent of the angle of friction between the two materials. (Rankine's Formula).

Following the above line of reasoning and omitting the mathematics, Marston's Formula for determining the load on an underground conduit is as follows:

Load on the underground trench conduit equals load coefficient for trench conduits times unit weight of the fill material times the square of the horizontal width of the trench at the top of the conduit or

$$Pe = Cd \times w \times \frac{Bd^2}{Bc}$$

$$Cd = \frac{1 - e^{-\alpha'H}}{2K\mu'}$$

$$\mu' = \frac{2K\mu'}{BD}$$

Where,

Pe = Vertical earth pressure (kg/m^2)

w = Unit weight of the fill material (kg/m^3)

Bd = Width of the trench (m)

Bc = Outside diameter of the pipe (m)

μ = Coefficient of internal friction of the fill material - tan $\varnothing$

μ' = Coefficient of sliding friction between the fill material and the sides of the trench = tan $\varnothing$'

$$K = \frac{\sqrt{\mu^2 + 1} - \mu}{\sqrt{\mu^2 + 1} + \mu}$$

H = height of the fill (m)

e = base of natural logarithm

μ and μ' depend on the type of fill material and on the type of soil in which the trench is dug. The diagram shows five curves giving values of Cd for values

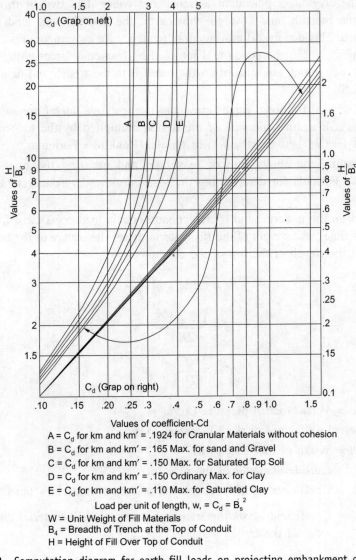

Values of coefficient-Cd

A = C_d for km and km' = .1924 for Cranular Materials without cohesion

B = C_d for km and km' = .165 Max. for sand and Gravel

C = C_d for km and km' = .150 Max. for Saturated Top Soil

D = C_d for km and km' = .150 Ordinary Max. for Clay

E = C_d for km and km' = .110 Max. for Saturated Clay

Load per unit of length, w, = $C_d = B_s^2$

W = Unit Weight of Fill Materials

B_4 = Breadth of Trench at the Top of Conduit

H = Height of Fill Over Top of Conduit

Fig. 8.3 Computation diagram for earth fill loads on projecting embankment conduits

of K and μ for five types of soil such as granular material without cohesion, sand and gravel, saturated top soil, ordinary clay and saturated clay.

Note: The width of the trench is taken as the horizontal width at the top of the conduit. If the trench has sloping side, the load on the pipe is equal to that for a vertical sided trench with a width equal to the width at the level of the top of the pipe.

8.3   EMBANKMENT CONDUITS

Embankment conduits are those that are covered by fills or embankments such as railway embankments, highway embankments and earth dams. Included also in this type are trench conduits installed in trenches wider than two or three times the outside diameter of the conduit. The typical example of an embankment conduit is the highway culvert.

Embankment conduits are divided into two classes:

8.4   PROJECTING CONDUITS

Projecting conduits are embankment conduits installed in shallow bedding with the top of the conduit projecting above the surface of the natural ground and then covered with an embankment. This type also includes those conduits which are installed in trenches wider than two or three times the maximum outside diameter of the conduit.

8.4.1   Negative Projecting Conduits

These are embankment conduits installed in relatively narrow trenches of such depth that the top of the conduits are below the level of the natural ground surface and then covered by a fill the height of which above the top of the conduit is appreciably greater than the distance from the natural ground surface to the top of the conduit.

8.4.2   Loads on Projecting Conduits

Loads on projecting embankment conduits involve other factors than those involved in the trench conduit condition or in the negative projecting conduit condition. Nevertheless, the basic principles of computing the loads are similar to those for computing the loads on trench conduits.

Some of the other factors involved in the projecting embankment conduit condition are as follows:

(1) **The Interior Prism** is the prism of backfill in the embankment directly over the conduit and between the vertical planes tangent to the side of the conduit.

(2) **The Exterior Prism** are those on each side of the conduit adjacent to the tangent vertical planes and of indefinite width.

There is a tendency for the exterior prisms in the projecting embankment condition (because they are higher) to settle more than the interior prism and for friction forces or shearing stresses to be exerted along the tangent vertical planes bounding the interior prism. These shearing stresses are transmitted to the conduit making the load on the structure greater than the weight of the interior prism of soil.

The magnitude of the shearing stresses (again neglecting cohesion) is assumed to be equal to the active lateral pressure at those planes times the coefficient of friction of the fill material.

In the case of higher fills, the shearing stresses will not extend to the surface but will terminate at some horizontal plane between the top of the conduit and the top of the embankment, known as the "Plane of Equal Settlement".

(3) **The Plane of Equal Settlement** is the horizontal plane in the embankment at and above which the settlements of the interior and exterior prisms of soil are equal.

(4) **The Height of Equal Settlement** is the distance between the top of the conduit and the plane of equal settlement.

(5) **The Critical Plane** is the horizontal plane in the fill material at the level of the top of the conduit at the beginning of construction of the embankment and before settlement have begun to develop.

The plane of equal settlement is influenced by two principal factors:

A. The settlement of the undisturbed sub-grade under the exterior prisms adjacent to the conduit and

B. The settlement of the top of the conduit.

When the critical plane settles more than the top of the conduit, the shearing stresses act downward on the interior prism, and when it settles less, they act upward.

A neutral case occurs when the top of the conduit settles downward an amount just equal to the settlement of the critical plane.

In this case the load on the conduit is just equal to the weight of the prism directly over it neither more than or less than that.

Projecting conduit foundations can be divided into two general cases.

CASE 1 where the conduit is on hard foundation.

CASE 2 where the conduit is on yielding foundation.

IN CASE 2 five factors may enter in:

1. Settlement of the conduit into its foundation.
2. Deformation of the conduit.

3. Settlement of the natural ground adjacent to the conduit.

4. Settlement of the "critical plane".

5. Compression of the columns of soil adjacent to the conduit, i.e., between the critical plane and natural ground.

(6) The "Settlement Ratio" is a semi-empirical abstract mathematical ratio combining all the above five factors and was determined on "a priori" reasoning and is not subject to mathematical proof. Thus, it might be considered as ASSUMPTIN NO 5. It is expressed in terms of the above as follows:

$$\text{Settlement ratio} = \frac{(5 + 3) - (1 + 2)}{5}$$

If the settlement ratio is positive, it means that the shearing stresses are additive or the critical plane settles more than the top of the conduit and vice versa.

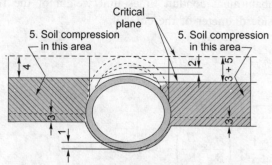

1. Settlement of the conduit into its Foundation
2. Deformation of the conduit
3. Settlement of the natural ground adjacent to the conduit
4. Settlement of the "Critical plane"
5. Compression of the columns of soil adjacent to
   the conduit, I.E. between the level of the orighinal critical plane
   and the final position of the natural ground

--- --- --- --- --- Before fill Material is placed above the level of the critical plane. H-O.

.... Position of the conduit before defformation.

——————————— Final position-after settlement

$$\text{Settlement ratio} = \frac{(5 + 3) - (1 + 2)}{5}$$

Fig. 8.4

(7) The Projection Ratio is the vertical distance from the natural ground to the critical plane divided by the overall width of the conduit. For convenience, field loading conditions have been grouped into the four following cases:

1. **"Complete Projecting Condition"** when the top of the conduit settles less than the critical plane and when the height of the embankment is less than the theoretical height of equal settlement.

2. **"Incomplete Projecting Condition"** when the top of the conduit settles less than the critical plane and when the height of the embankment is greater than the height of equal settlement.

3. **"Complete Trench Condition"** when the top of the conduit settles more than the critical plane and when the height of the embankment is less than the theoretical height of equal settlement.

4. **"Incomplete Trench Condition"** when the top of the conduit settles more than the critical plane and when the height of the embankment is greater than the height of equal settlement.

Load on the projecting conduit equals the load coefficient for the projecting embankment conduit times unit weight of the fill material times the outside diameter of the conduit.

or

$$P_e = C_c \, W \, Bc \qquad \qquad ...(4.1)$$

where,

$$C_c = \frac{e^{\beta H} - 1}{2k\mu} \text{ (when } H \le H_e)$$

$$C_C = \frac{e^{\beta h} - 1}{2k\mu} + \left(\frac{H}{B} - \frac{H_e}{B_C}\right) \cdot e^{\beta He}, \text{ (when } H > H_e)$$

$$\beta = \frac{2K\mu}{Bc}$$

The value of He may be obtained form the following relations.

$$e^{\beta He} - 2K\mu \frac{He}{Bc} = 2K\delta \cdot p' + 1$$

Where,

He = Height of equal settlement
δ = Settlement ratio
P′ = Projection ratio

Values of He and C_c are shown in Table 8.1, at the case of
P′ = 0.7 and P′ = 0.5 respectively.

Table 8.1 Values of He and Cs (Projecting condition)

H ≤ He	H > He	
	δ·p′ = 0.7	δ·p′ = 0.5
$C_c = \dfrac{e^{0.384H/Bc-1}}{0.3848}$	He = 1.70 Bc	He = 1.46 Bc
	C_C = 1.924 H/Bc − 0.869	C_C = 1.754H/Bc − 0.602

8.4.3 Loads on Negative Projecting Embankment Conduits

A negative projecting embankment conduit is a type of installation which falls in between the trench conduit and the projecting conduit. Negative Projecting Embankment Conduits are conduits constructed in relatively narrow trenches of such depth that the top of the conduit is below the level of the natural ground surface and covered by a fill, the height of which above the top of the conduit is appreciably greater than the distance from the ground surface to the top of the conduit. An important variation of this is the "imperfect trench" which is an artificial negative projecting installation.

Culverts are sometimes located in higher ground to one side of the natural water course and the stream flow is diverted to the new location. The culvert pipe may than be laid in a relatively narrow trench dug in the undisturbed soil of the side hill. If the depth of the trench is such that the top of the pipe is below the natural ground, the culvert is considered to be a Negative Projecting Embankment Culvert. The load transmitted to the pipe equals the weight of the interior prism of soil above the pipe plus or minus friction forces along the side of that prism as in the case for the Projecting Conduit, however, the width of the interior prism is determined by the trench width rather than by the outside diameter of the pipe.

The procedure for computing the load on a negative projecting culvert pipe is somewhat similar to that used in computing the load on a positive projecting conduit. However, it should be remembered that a Negative Projecting Conduit has some the characteristics of a trench conduit even through it is under a fill. The formula for computing the load on a Negative Projecting Conduit is:

$$P_e = Cn \, w \, \frac{Bd^2}{Bc}$$

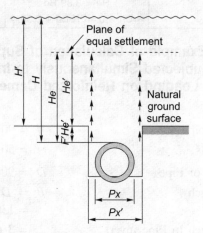

Fig. 8.5 Negative projecting conduit

$$C_n = \frac{1 - e^{-\alpha \cdot H}}{2K\mu} \quad \text{(when } H \leq H_e)$$

$$C_n = \frac{1 - e^{-\alpha \cdot He}}{2K\mu} + \left(\frac{H}{Bd} - \frac{He}{Bd}\right) - e^{-\alpha \cdot He} \quad \text{(when } H > H_e)$$

$$\alpha = \frac{2K\mu}{Bd}$$

The values of He may be obtained from the following relations.

$$= \left\{\left(\frac{H'}{Bd} + \frac{He}{Bd}\right)\right\} - \frac{1}{2k\mu} \frac{1 - e^{-\alpha \cdot He'}}{2H\mu} - \left\{\frac{He'}{Bd} - \left(\frac{H'}{Bd} - \frac{He'}{Bd}\right) \frac{1He'}{2Bd} - \frac{1}{2k\mu}\right\}$$

$$= \left\{\frac{2}{3}\delta p' \frac{1 - e^{-\alpha \cdot He'}}{2K\mu} + \left(\frac{H'}{Bd} - \frac{He'}{Bd}\right) e^{-\alpha \cdot He'}\right\}$$

Where, $H = H' + p' \, Bd$, $He = He' + p' \, Bd$.

Values of He and Cn are show in Table 8.2, at the case of

$\delta = -0.3$, $K\mu = 0.130$ respectively.

Table 8.2 Values of He and Cn (Negative projecting condition)

$H \leq He$	$H > He$	
	$p' = 0.5$	$p' = 1.0$
$C_n = \dfrac{1 - e^{-0.260 \, H/Bd}}{0.260}$	He = 2.00 Bd	He = 3.03 Bd
	Cn = 0.71 H/Bd + 0.14	Cn = 0.58 H/Bd + 0.34
	$p' = 1.5$	$p' = 2.0$
	He = 3.89 Bd	He = 4.82 Bd
	Cn = 0.48 H/Bd + 0.58	Cn = 0.40 H/Bd + 0.82

8.4.4 Typical Example of Calculation of Supporting Strength of a Pipe Subjected Simultaneously to Internal Pressure and External Loadind on Reinforced Cement Concrete Pipes

Data

Diameter of pipe	$d = 900$ mm
Wall thickness	$t = 50$ mm
External dia. of Pipe	$D = 900 + 2 \times 50 = 1000$ mm
Width of trench	$B = D + 300 = 1000 + 300$ $= 1300$ mm $= 1.3$ m
Depth of trench to Pipe invert	$= 3$ m
Weight of fill material	$W = 17.167$ KN/m^3

Load Factor $Ft = 1.5$

Depth of fill material over top of Pipe $H = 3 - 1.3 = 1.7$ m

Internal Working Pressure $Pw = 0.06867$ N/mm^2

Hydraulic Test Pressure at Factory $Pt = 0.13734$ N/mm^2

Formula as per IS : 783 : 1985 clause 13, Page No 20

Internal Pressure

$$T = \frac{W}{F} \frac{(Pt)^{13}}{(Pt - Pw)}$$

Three edge bearing load when simultaneous

Internal pressure and three edge pressure at failure.

Where,

W = Site external working load in KN/m of pipe.

F = Load Factor

T = External three edge bearing load per meter pipe,

Pt = Hydraulic test pressure at factory in MPa,

Pw = Working pressure on the line in MPa, and

$\dfrac{W}{b}$ = Test load equivalent to the site external working load W.

Calculations

Vertical Load on pipe in trench due to fill Material

$$We = Ct \times Ke \times B^2$$

Value of Ct is obtained from Fig. 8.4.

$$\frac{H}{B} = 1.31$$

Clay sand $Ct = 1.16$

Calculated earth load on top $= We = Ct \times Ke \times B^2$

$$= 1.16 \times 17.167 \times 1.3^2$$

$$= 33.654 \text{ KN/m}$$

Load Factor $= Ft = 1.5$

Load to produce 0.25 mm crack $= We/Ft = 33.654/1.5$

$$= 22.44 \text{ KN/m}$$

For simultaneous internal pressure and external pressure

$$T = \frac{We}{F} \frac{(Pt)^{1/3}}{(Pt - Pw)}$$

$$T = \frac{33.654}{1.5} \frac{(0.14)}{(0.14 - 0.07)}$$

$$T = 28.27 \text{ KN/m}$$

Hence P2 class pipe with NP2 class 0.25 mm pipe load is to be used NP2 class load is 28.27 KN/m.

TYPICAL EXAMPLE OF CALCULATION OF LOAD ON PIPE AND SELECTION OF CLASS – TRENCH CONDITION

Data

Dia. of pipe	d	= 900 mm
Wall thickness	t	= 50 mm
Depth of trench to pipe invert		= 2.93 meter
Density of fill material		= 1750 kg/m^3 = 17.167 KN/m^2
Characteristic of filling material		
Ku & Ku′		= 0.13 (Wet clay)
Pipe bedding method		= Ordinary bedding
O.D. of pipe BC		= $d + 2t$
		= 900 + 100
		= 1000 mm

Height of fill over top of the pipe = H = 2.93 – 1

1.93 meter

Width of trench $Bt = Bc + 300$ (Assume)

$$= 1 + 0.3$$
$$= 1.3$$

Load factor   = 1.5

Calculation

The vertical load w on the pipe
due to the fill material $W = Ct \ W \ Bt^2$

Ct = Coefficient to be taken from I.S. 783-1985

(This depend upon the ratio of H/Bt and the characteristic of filling material)

$$\frac{H}{Bt} = \frac{1.93}{1.3} = 1.45$$

Ct from graph   = 1.25

Calculated load on pipe   = 1.25 × 1750 × 1.3^2

per meter length   = 3720 kg

Three edge bearing load $\qquad = \dfrac{3720}{1.5}$

With load factor of 1.5 $\qquad$ = 2490 kg

Use NP2 class pipe three edge bearing load for 0.25 mm crack is 2500 kg.

TYPICAL EXAMPLE OF CALCULATION OF LOAD ON PIPE AND SELECTION OF CLASS – EMBANKMENT CONDITION

Data

Dia. of pipe	d = 900 mm	
Wall thickness	t = 50 mm	
Depth of filling		
to pipe invert	= 2.90 meter	
Density of filling		
material	= 1750 kg/m^3	
characteristic of filling		
material Ku	= 0.13 (Wet clay)	
Pipe bedding material	= Ordinary bedding	
O.D. of pipe	BC $= d + 2t$	
	= 900 + 100	
	= 1000 mm	
Height of fill over		
top of the pipe	H = 2.9 − 1.0	
	= 1.90 meters	
	Load factor = 1.95	

Calculation

The vertical load on the
pipe due to fill material $W = Ct\ W\ Bc^2$

Ce = Coefficient to be taken from I.S. 783 – 1985

(This depend upon H/Bc, settlement ratio and projection ratio. Assume settlement ratio as 0.7 and projection as 0.7)

$$= \frac{H}{Bc} = \frac{1.9}{1.0}$$

$$= Ct = 2.75$$

$$We = 2.75 \times 1750 \times 1^2$$

$$= 4812.50 \text{ kgs}$$

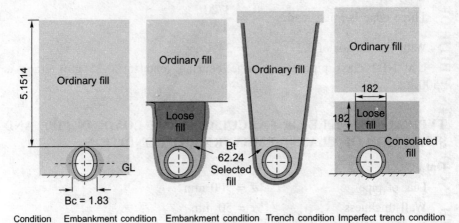

Condition	Embankment condition Positive projection	Embankment condition Negative projection	Trench condition	Imperfect trench condition
Load factor	482.75 kg/m	28610 kg/m	.24585 kg/m	
Three edge	1.95	1.5	1.5	1.95
Bearing load	24755 kg/m	19075 kg/m	16390 kg/m	9900 kg/m

Data

1. ID of pipe	= 1.52 m	4. Bedding	= Ordinary	
2. Od of pipe	= 1.83 m	5. Density of filling material	= 1925 kg/m³	
3. Trench width	= 2.44 m	6. Projection ratio	= 0.7	
7. Settlement ratio	= 0.5			

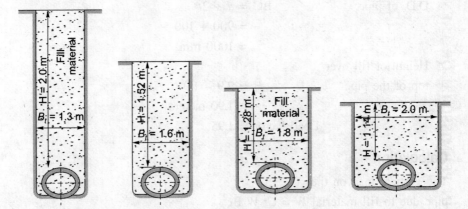

Fig. 8.6 Ordinary bedding

Three edge bearing load

With a load factor of $\quad 1.95 = \dfrac{4812.5}{1.95}$

$$= 2467.94 \text{ kgs}$$

Use NP2 class pipe. Three edge bearing load to produce 0.25 mm crack is 2500 kgs.

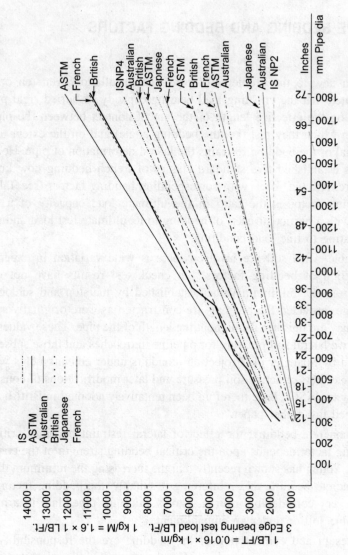

TYPICAL VARIATION IN PERMISSIBLE OF TRENCH CONDUCT WITH INCREASE IN WIDTH OF TRENCH

Fig. 8.7 3EDGE bearing loads for various diameter as as per diefferent specifications

Data

ID OF PIPE = 900 mm

OD OF PIPE = 1000 mm

DENSITY OF FILL MATERIAL = 1750 kgs/cm^2

$K\mu$ =0.13

8.5 PIPE BEDDING AND BEDDING FACTORS

General

If we turn now to the problems of pipe line installation, Marston originally demonstrated that the crushing load bearing capacity of buried rigid pipe line depended on the effective length of the arc of contact between the pipes and the bed on which they rest (i.e. the bedding angle) and cn the extent to which the material of the bedding restrains the lateral deformation of pipe. He and his colleagues established five standard classes of trench bedding now known as classes Arc, A,B,C,D each with corresponding bedding factors (see Table 8.1). The bedding factor is the ratio of the ultimate load capacity of a pipe as installed with a particular class of bedding to its ultimate test load and reaction approximating to line load condition.

Inevitable when soil are involved there is wide variation in experimental determination of bedding factors, but check test results have not differed sensibly from the minimum values established by marston and schlock some 40 year ago, provided the bedding are constructed as were originally specified and that they are uniform throughout the length of the pipe. These values, which ignore active lateral soil pressure for pipeline in trenches and those subsequently established by Spangler of 'Projecting' conduits under embankments which do include the effects of active soil pressure and later modifications of construction adopted by the ASCE have therefore been tentatively adopted in British practice as described in another paper.

With concrete bedding, the effect of lateral restraint becomes critical and the bedding factor depends upon the critical bedding strength of the concrete at the invert. Young has shown recently that the increasing the minimum thickness of the concrete bedding and reinforcing it suitably, the bedding factor can be increased very considerably over marston's value for pipes laid in a trench in compressible soil or under an embankment.

The design and construction of the bedding are the responsibility of the Engineer and contractor respectively. But the pipe maker is interested to know that their contribution to the strength of the finished pipe line properly and efficiently complements the guaranteed crushing strength of his pipe. Frequently his advice is also needed and is sought, where alternative pipe strengths have to be considered or in peculiar or difficult soil or loading condition. Early co operation between the designer and the pipe maker can therefore be of great help and value in the author's experience, even with the traditional method of pipe laying faulty design or construction rather than to faulty pipes. This situation is likely to be considerable improved with the increasing use of extra strength pipes, but it will not be cured unless and until Engineer and contractors appreciate their responsibility and take steps to instruct and train their personnel in the essential principles and correct technique of pipe installation. At present, inspectors, foremen and pipe layers are generally insufficiently aware either of these principles or of their responsibility for pipeline strength and for avoidance of unnecessary load and forces acting on the pipes. Then, since the pipe maker is called upon to guarantee the minimum crushing strength of his pipes, the contractor should equally be called upon to guarantee a specified minimum bedding factor and its uniformity throughout the length of the pipeline.

Bedding design

It will be appreciated that in any pipe line the required load carrying capacity of the pipes may be achieved by utilizing either weak pipes with a strong bedding or stronger pipes with a weaker bedding. This choice has been made possible by the recent introduction of the so called 'Extra Strength' (i.e. stander strength graded) pipe by British stander 556. formerly there was no alternative to the concrete bedding or "support" as it was called in the form of either bedding and hunching, or a complete surround, neither of which was reinforced and neither of which had an established bedding factor. Numerous failures in the field and laboratory tests indicated that with concrete of unspecified strength placed in unspecified conditions and of arbitrary thickness it was impossible to establish a bedding factor and the load capacity of pipe line was therefore a matter of chance. Even with a fully specified bedding of marston Arc or A class (bedding factors 3.4 or 2.6 respectively) and an established bedding factor, the placing and curing of site concrete is exacting, slow and very dependent upon weather conditions, and complications occur with flexible joints and delay necessitated by the curing and hardening of the concrete.

For all these reasons the use of higher strength pipes with the class B granular bad (bedding factor 1.9) is to be preferred under normal conditions as being quicker to construct in any weather conditions, including frost, provided the granular bedding material is dry and free from i.e., Exceptionally, in bed ground where for any reason a granular bad could be disturbed, a concrete bed may be essential. Normally, however, the concrete bad should be regarded as the exception rather than the rule as formerly.

Practical factors affecting bedding

The inferior bedding classes C and D (bedding factor 1.5 and 1.1 respectively) may exceptionally be used in uniform soils but in mixed soils, such as frequently occur in this country, they suffer from the common fault that properties of the soil are too variable to provide the essential uniformity of support to the barrels of the pipes. This succession of hard and soft zones induces variable reaction along the pipe axis with consequent axial bedding stresses and transverse shear stresses, either of which may fracture the pipe.

In the class B granular bed, these irregularities and variations can be largely if not entirely eliminated, and both axial and transverse uniformity of support ensured by the proper selection of bedding materials and proper care in its placing and uniform compaction. Without this care in construction the bedding factor may be reduced to that of a class C or even of a class D bed with the added pipe stresses associated therewith.

Attention should also be paid to any lack of uniformity in the foundation underlying the bedding weak spots being strengthened and hard spots eliminated the objective being, as in the bedding to make the foundation as uniform as possible throughout its length. Sudden changes from rock to softer material, especially at shallow level, require special attention and precautions during pipe laying irrespective of the type of bedding used, as otherwise the pipes may be fracture when subjected to high concentrated surcharges.

A possible source of weakness in a class B granular bed is the subsequent moment of bedding material caused by the withdrawal of side sheeting, runners or pipes after the bedding has been placed, thus leaving voids into which the bedding material can flow so reducing or disturbing its support of the pipes. This effect obviously becomes worse as the width of the "timber" increases and that of the trench decreases. The withdrawal of sheeting after back filling also destroys the frictional forces in the trench wall and so increases the load on the pipe in a narrow "trench".

The essential requirements of granular bedding materials have been discussed in another paper and very useful observations on grading and stability have been made recently by storey after extensive site experience.

Future objectives

Future problems requiring investigation are the actual effects produced by faulty bedding; the sinking of pipes into a granular bed and the effects on pipe alignment of the swelling and shrinking of a clay foundation when the trench is excavated and backfilled respectively.

Wheel load and impact factors

The high dynamic loading to which roads, railways, airfields or other traffic bearing structure are subjected mainly concerns the designers of such structures and/or of any pipe lines laid beneath them. Pipe makers are also concerned, however, in deciding upon the upper limit of the crushing test strength of large pipes which may be needed for such conditions. If the maximum wheel loads and their spacing are specified, the maximum static load imposed upon a pipeline will depend upon the numbers of wheels which can influence it simultaneously wheel group relative to the pipe line and it can be computed.

The factor by which this maximum static load or any of its components must be multiplied to make it equivalent to the dynamic load which is imposed on the pipeline when any wheel or group of wheels passes over a surface irregularity is know as the "impact factor". Although such factors are frequently specified by road or other transport authorities, they are not well established or specified in sufficient detail, as is evident from Table 8.2.

The absence of recognized standard test conditions most probably accounts for the wide variation in the factors used by different authorities, since some or all of the following conditions may affect the value of an 'impact factor' depth of cover, degree of roughness of the surface, speed of the vehicles, resilience of the materials between the pipes and the surface, flexibility of the pipe bedding and joints; length and diameter of the pipes, type of tires and suspension of the vehicle and whether or not it is liable to be overloaded; and the possibility of temporary or permanent changes in any of these conditions with time, wear and tear, climatic conditions, maintenance operations, new works, etc.

This situation is recognized by all concerned as far from satisfactory and an active research program has been initiated in this country by the authorities responsible for transport and for local Government with the object of obtaining a more representative value of the factor or factors and realistic approach to their application to roads in particular. The concrete pipe association has also initiated research on the behavior of concrete pipes under high transitory and repeated loading.

In view of current uncertainties, and on the advice of the ministry of transport road research Laboratory, on the basis of their researches on road beds, an impact factor of not less than 2.0 at all depths and on any type of road has been tentatively adopted for great Britain. This factor is applied to one of the two axles, eight wheel groups of the maximum current bridge loading specified as type HB in BS 153 with a static load of 45 tons per axle.

Technical administration

Much loss of time due to delay in the delivery of large pipes which are not stock items could be avoided if the order for the pipes could be placed by the Engineer some time in advance of the letting of contract for the installation. This is contrary to the usual practice in the U.K. and would need safeguards regarding delivery, storage, responsibility for damage, and so on. The author believes however, that it would also help the Engineer by enabling him to adjust his factor of safety by varying the bedding class to suit the variable skill and experience of the chosen contractor, and so produce a more realistic comparison of tenders and eventually a more efficient job.

Conclusion

It is hoped that this brief view has indicated the present liveliness of the industry and its readiness to take its place in the general technological advance of the age. The willingness and even enthusiasm show by members of the industry in general and by the pipe makes in particular, to reconsider old problems in the light of a more rational approach, and their desire to probe further, augur well. Great changes have been introduced in the past ten years, and more on the way. This interest in and respect for a new and lifted known specialist technology is most welcome and largely coincides with similar developments in most other industrialized countries. The new methods have been accepted in principle by the ministry concerned with sewerage and by the drafting panels revising the national civil engineering code of practice on sewerage and the building code of practice on house drainage.

The foundation of the new technique has been laid. Further developments and refinements may be expected as the result of current investigations both here and abroad and later, it is hoped, by closer attention to the many soil mechanics problems involved in both load estimation and pipe line installation technique.

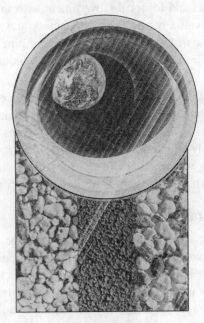

CHAPTER 9

Design of Concrete Pipe

9.1 INTRODUCTION

During the last decade the designing of rigid underground pipes and their joints and the technique of their installation have been changing fundamentally from empirical and traditional method to more rational methods based on the scientific principles of structural and soil mechanics. A buried pipeline is now recognized as a load bearing structure, and within the limits of present knowledge, is designed with growing confidence. The essential and important design change in the past decade is that the primary gravitational loads imposed upon a pipeline by the soil overlying it and by surface surcharges, either static or dynamic, are estimated by the methods developed by Marston, Spangler and Schlick in the USA and equated to the known and tested minimum crushing ultimate or cracking strength of the pipes as modified by a known bedding factor and an assessed safety factor to cover variations in site work.

9.2 DESIGN OF CONCRETE PIPE FOR EXTERNAL LOAD

9.2.1 Reinforced Concrete Pipes

A simple structural member like RCC Pipe should not be difficult to design in normal course, but it is; because of certain inherent consideration involved in it. Even today, there is no uniformity in design of RCC Pipes in different countries. Most of the specifications, like EN, Indian specification and others are silent about the design of RCC pipe. The main reason is that, it is based on crack control concept.

When the concrete pipe was first made, there was a problem as to how to test it? It was then thought that a load should be applied on the top of pipe with line support at the bottom. The load should be, increased till a crack is observed at the crown from inside (the three edge bearing test). Next problem was how to define the crack? Then it was agreed that 0.25 mm wide crack for a length of 300 mm will give the required load. Next problem was how to measure the crack? Then it was considered that a feeler gauge of 0.127 mm thickness should pass though it. When all these were accepted; considerable work was done during last 40-50 years throughout the world to find out the stress in steel at those conditions. After doing considerable work and testing it on pipes, Humes of Australia developed two formulae for the design. The formulae do considered allowable tensile stress in bending for concrete, thickness of concrete and the lever arm for steel. These formulae are given below.

$$(1) \quad fs = 2465 \sqrt{\text{Krup} \; \frac{\Delta}{S} \; \frac{(1-k)}{(t/d - k)}}$$

$$(2) \quad \text{As} = \frac{\text{B.M}}{0.875 \; d \times fs}$$

Where,

fs = Stress in steel in kg/cm^2

Krup = Allowable tensile stress in concrete due to bending in kg/cm^2

Δ = Actual crack width into which a filler gauge can be inserted

= 0.3045 mm

s = Diameter of spiral wire in mm

t = Wall thickness of pipe in mm

d = Depth of steel from extreme fiber

As = Area of steel in cm^2

k = Constant value = 0.3

c = Clear cover

9.2.2 Bending Moment in the Pipe Wall

Bending moments will be induced in the wall of buried pipes which are subjected to external loads. In addition to the external loads, weight of pipe itself, weight of water in the pipe and lateral earth pressure are also the causes to induce moments in the pipe wall. In general conditions of pipe laying, amount of moments due to pipe weight and water weight counter balances to that caused by lateral earth pressure.

Although, at computing the moment, vertical component is solely taken in consideration, conditions of pipe bedding and of foundation, characteristics of surrounding soil and the method of pipe laying affect greatly to the amount of moment, and sufficient care should be paid to these factors.

In R.C.C. pipe, tension at crown is inside while at springline, it is outside.

The B.M at springline is half of B.M at crown. Hence, for design purpose, the area at springline should be half that of crown, but because lever arm is small there, area of the steel at springline = 0.75 × area of steel at crown. For different bedding angles, the B.M is different.

(1) Bending Moment

For concentrated load in the centre for a freely supported beam

B.M = 0.25 PL

Where, P = Load,

   L = Span.

In the case of pipe, the beam is curved hence,

B.M = 0.159 P Dm

Where, Dm = Mean diameter of pipe

(2) Area of Steel As

$$As = \frac{B.M.}{j \times d \times fs} \text{ cm}^2/\text{m.} = \frac{0.159 \; P \; Dm}{j \times d \times fs}$$

Where, j = Lever arm

   d = Effective depth of steel

(3) Wt. of Steel per meter, Wt.

$$Wt = \frac{As \times \pi \times (D + 2c)}{\text{Area of wire in cm}^2} \times \text{Weight of wire per mtr.}$$

Where,

   D = Internal diameter of pipe in meter

   c = Clear cover in meter

   W = Weight of Spiral Steel per meter

9.3 TYPICAL DESIGN OF R.C.C. PIPE - ILLUSTRATIVE EXAMPLE NO 1

Data

Internal diameter of pipe,	D = 900 mm NP^2	Class	
Wall thickness,	t = 55 mm	= 5.5 cm	
Inside clear cover,	c = 15 mm	= 1.5 cm	
Dia. of spiral wire,	S = 4 mm		
Three edge bearing load	p = 22.8 kN/meter		
Tensile strength of concrete in bending	Krup = 50 kg/cm^2 for Concrete strength of 350 kg/cm^2		

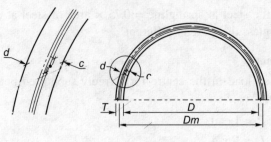

Fig. 9.1

Calculation

Wt. of 4 mm dia. wire = 0.09865 kg/meter

$$d = 55 - 15 - 4 - 2 = 34 \text{ mm or 3.4 cm}$$

$$t/d = 5.5/3.4 = 1.62$$

Stress in steel $fs = 2465 \times \sqrt{\dfrac{\text{Krup } \Delta(1 - K)}{S(t/d - K)}}$

$$= 2465 \times \sqrt{\frac{50 \times 0.3048(1 - 0.3)}{4 \times (1.62 - 0.3)}}$$

$$= 2465 \times \sqrt{\frac{50 \times 0.3048 \times 0.7}{4 \times 1.32}} = 2465 \times \sqrt{\frac{15.24 \times 0.7}{4 \times 1.32}}$$

$$= 2465 \times \sqrt{2.04} = 2465 \times 1.43 = 3443 \text{ kg/cm}^2$$

Area of steel $As = \dfrac{0.159 \times 22.8 \times 101.972 \times 95.5}{0.875 \times 3.4 \times 3443} = 3.55 \text{ cm}^2$

Wt. of steel $Wt = \dfrac{3.55 \times 3.1416(900 + 55)}{0.1257 \; 100} \times 0.09868 = 8.27 \text{ kg/m}$

In this case the cage is one but when thickness is more than 80 mm double cage has to be used.

9.4 TYPICAL DESIGN OF R.C.C. PIPE - ILLUSTRATIVE EXAMPLE NO 2

Data

Internal diameter of pipe	D = 1200 mm	NP3 Class
Wall thickness	t = 120 mm	= 12.0 cm

Inside clear cover $\qquad c$ = 15 mm = 1.5 cm

Dia. of spiral wire $\qquad S$ = 6 mm

Dia. of longitudinal wire $\qquad$ = 6 mm

Three edge bearing load $\qquad p$ = 22.8 kN/meter

Tensile strength of concrete $\quad$ Krup = 50 kg/cm^2 for Concrete strength
in bending $\qquad\qquad\qquad$ of 350 kg/cm^2

Calculation

Wt. of 6 mm dia. were = 0.222 kg/meter

$$d = 120 - 15 - 6 - 3 = 96 \text{ mm or } 9.6 \text{ cm}$$

$$t/d = 12/9.6 = 1.25$$

$$\text{Stress in steel } fs = 2465 \times \sqrt{\frac{\text{Krup } \Delta(1 - K)}{S \,(t/d - K)}}$$

$$= 2465 \times \sqrt{\frac{50 \times 0.3048(1 - 0.3)}{4 \times (1.25 - 0.3)}}$$

$$= 2465 \times \sqrt{\frac{50 \times 0.3048 \times 0.7}{4 \times 0.95}} = 3372.26 \text{ kg/cm}^2$$

$$\text{Area of steel } As = \frac{0.159 \times 57.48 \times 101.972 \times 132}{0.875 \times 9.6 \times 3372.26} = 4.34 \text{ cm}^2$$

$$\text{Wt. of steel } Wt = \frac{4.34 \times 3.1416 \,(1200 + 120)}{0.2827 \times 1000} \times 0.222 = 14.14 \text{ kg/m}$$

As thickness is more than 80 mm double cage is to be used.

Spiral wt. of steel for inner cage $\quad$ = 14.14 kg/m

Spiral wt. of steel for outer cage $\quad$ = 14.14 × 0.75 $\quad$ = 10.60 kg/m

Total spiral wt. of steel $\qquad\qquad$ = 14.14 + 10.60 $\quad$ = 24.74 kg/mtr

9.5 UNREINFORCED CONCRETE PIPES

Introduction

Unreinforced concrete pipes have a number of technical and economical advantages compared to reinforced concrete pipes. For example, the risk of corrosion of the reinforcement leading to disintegration of the pipe is completely avoided for unreinforced concrete pipes in sewage systems. Furthermore, the cost of the reinforcement for each pipe is saved and the production facility for reinforcements is not needed, leading to savings in establishment cost of the concrete pipe manufacturing facility. Another economical advantage is that higher production rates can be achieved, since time consuming handling of reinforcement is avoided. Due to these advantages production of unreinforced concrete pipes has been taken up, by a large number of pipe manufactures in countries like UK, USA, Ireland, Norway, Denmark, Sweden, Germany etc. (reference list is enclosed). Furthermore, the governments in these countries have accepted unreinforced concrete pipes and the concrete pipe standards have been adopted to accommodate unreinforced pipes.

The standards for unreinforced concrete pipes in these countries are all slightly different with respect to load requirements, wall thickness, concrete strength, size limits, design guidelines etc. However, the bases for all the standards are static analysis of the two failure modes for concrete pipes as shown in Fig. 9.2.

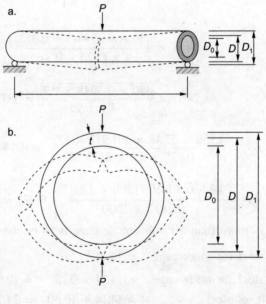

Fig. 9.2

The failure modes shown are called beam failure and crushing failure respectively. Beam failure is typically the critical failure mode for pipes with internal diameter less than 400–500 mm, depending on the pipe length. Crushing failure is the critical failure mode for larger diameter pipes. However, both failure modes must be considered when calculating the minimum wall thickness in order to comply with the standards.

9.6 DESIGN OF UNREINFORCED CONCRETE PIPES

Germany is one of the leading countries in the world, in the production of large diameter unreinforced concrete pipes. The German concrete pipe association (Fachvereinigung Betonrohre and Stallbetonrohre e. V.) FBS may be the leading technical authority in the world in the design and production of unreinforced concrete pipes. Their technical knowledge is based on many years of practical experience from the German sewage infrastructure industry. It is therefore natural to use the German standards and codes as a base for the design of unreinforced concrete pipes for the Indian standard.

In the German standards, the load requirements are based on the crushing failure. Unreinforced concrete pipes can achieve an FBS approval if they pass a crushing test (three-edge hearing test) with the line loads per meter pipe length given in the figure below. For reference the loads requirements in the Indian standard IS 458 pipe class NP3 and NP4 is also given.

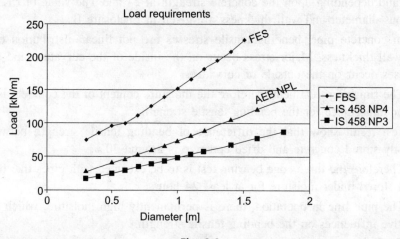

Fig. 9.3

The ultimate load requirement in the Indian standard for NP4 is only around 50% of the German FBS requirements. Therefore, a smaller wall thickness is sufficient to accommodate the loading given in Indian standard.

The calculations of the wall thickness for the Indian pipes are done in accordance with the guidelines described in DIN 4032. These guidelines are based on the linear elastic theory and use the tensile stress in the pipe wall under the three-edge bearing test as the critical parameter.

Bending moments

The bending moments sustained by the pipe are determined by the bending tensile stress sustainable by concrete. The bending tensile stress of the concrete is a strongly scattered value and depends upon various influences. The most essential once, are concrete quality, arrangement of loading, wall thickness, and influences of curvature and influence of fining treatment of concrete.

With growing concrete quality the bending tensile stress also increase. The following approximate formula may be given,

$$\sigma_{BZ} = 0.478^3 \sqrt{\beta^2_{w28}} = \text{Bending tensile stress}$$

$$\sigma_{BZ} = \text{Bending tensile strength of pipe}$$

$$\beta_{W28} = \text{Compressive strength of concrete at 28 days.}$$

For concrete strength of 45 N/m^2.

$$\sigma_{BZ} = 0.478^3 \sqrt{45^2} = 5.9 \text{ kp/cm}^2.$$

The tensile stress inside the crown of pipe is more and this is obtained by multiplying the above by a factor α K, Concrete bending tensile strength σ_{BZ} and depending upon the concrete strength at 28 and The value of σK for various diameter and wall thickness are given in Enclosure B.

In concrete pipe, bending tensile stresses are not linear distribution over the wall thickness, Major stress occur on the inside of the curvature and less stresses occur on the outside of curvature.

The finishing treatment therefore the moisture content of the concrete is of major importance for the bending tensile strength.

Test result show that the difference of bending tensile strength between moistly stored concrete and dried concrete is around 30%.

Therefore the three edge bearing test is to be executed with pipes that have been stored under moisture for at least 24 hours.

The pipe line in operation, there is permanently high moisture, which has positive influences on the bending tensile strength.

Furthermore, consideration must be given to special issues related to the vibration vertical casting technology. It is very important that the pipes are stable since the pipes are demoulded immediately after casting. Pedershaab A/S recommends a minimum wall thickness of 50 mm for a 300 mm diameter and 2500 mm long pipe.

Illustration Example is given below:

For design purpose bending moment of the pipe due to self weight of *G* in taken unit account in addition to the moment detect this edge bearing load.

The stress to be consider are,

MG = Stress at crown due self weight

= $0.07 \times G \times rm$

The stress due to the load in

$$MF = 0.3 \times F \times rm$$

The bending moment at crown causes a ring tensile stress according of

$$\sigma_{BZR} = \frac{\Sigma M}{W}$$

σ_{BZR} = Ring bending tensile stress.

W is The moment of inertia.

9.6.1 Illustrative Example No 1 of Unreinforced Concrete Pipe 1

Data

Internal diameter of pipe	D	= 900 mm
Load NP_3 class	F	= 43.11 kN/m
Wall thickness	T	= 100 mm

Calculation

Mean diameter of pipe $\quad\quad d = D + T = 900 + 100$

$$= 1000 \text{ mm} = 1.0 \text{ m.}$$

Radius $\quad\quad\quad\quad\quad\quad rm = \dfrac{D+T}{2} = \dfrac{1000}{2} = 500 \text{ mm} = 0.5 \text{ m.}$

Wt. of pipe/mtr $\quad\quad\quad G = \pi \times d \times 2500 = 3.14 \times 1.0 \times 2500$

$$= 7.85 \text{ MT/mtr}$$

Bending moments

Due to three edge bearing test $MF = 0.3 \times F \times rm$

$$= 0.3 \times 43.11 \times 0.5 = 6.467 \text{ kN/mtr}$$

Due to self wt. $\quad\quad\quad MG = 0.07 \times G \times rm$

$$= 0.07 \times 7.85 \times 0.5 = 0.2765 \text{ kN/mtr}$$

$\Sigma M = MF + MG = 6.467 + 0.2765 = 6.74 \text{ kN/mtr.}$

Moment of resistance W = Section modulus $= \dfrac{LT^2}{6}$

$$= \frac{1}{6} \times 100 \times 100 = 1666.67 \text{ cm}^3$$

Bending tension stress $\sigma_{BZR} = \dfrac{\Sigma M \times 10^3}{W} = \dfrac{6740}{1666.67} = 4.05$

Maximum bending tension inside crown $\quad = 4.05 \times 1.073 = 4.35$

9.6.2 Illustrative Example No 2 of Unreinforced Concrete Pipe 2

Data

Internal diameter of pipe	D = 1200 mm
Load NP_3 class	F = 88.30 kN/m

Wall thickness $\qquad$ $T = 135$ mm

Calculation

Mean diameter of pipe $d = D + T = 1200 + 135 = 1335$ mm $= 1.335$ m.

Radius $\qquad rm = \dfrac{D + T}{2} = \dfrac{1335}{2} = 667.5$ mm $= 0.6675$ m.

Wt. of pipe/mtr $\qquad G = \pi \times d \times 2500 = 3.14 \times 1.335 \times 2500$
$\qquad\qquad\qquad = 14.15$ MT/mtr

Bending moments

Due to three edge bearing test

$$MF = 0.3 \times F \times rm$$
$$= 0.3 \times 88.30 \times 0.6675 \qquad = 17.68 \text{ kN/mtr}$$

Due to self wt $\qquad MG = 0.07 \times G \times rm$
$$= 0.07 \times 14.15 \times 0.6675 \quad = 0.66 \text{ kN/mtr}$$

$\Sigma M = MF + MG = 17.68 + 0.66 = 18.34$ kN/mtr.

Moment of resistance W = Section modulus $= \dfrac{LT^2}{6}$

$$= \frac{1}{6} \times 135 \times 135 = 3037.5 \text{ cm}^3$$

Bending tension stress $\sigma_{BZR} = \dfrac{\Sigma M \times 10^3}{W} = \dfrac{18340}{3037.5} = 6.03$

Maximum bending tension inside crown $= 6.03 \times 1.065 = 6.42$

Indian Scenario - I.S. 458

Spiral steel for spinning process is higher than spiral steel for vertical cast process for same three edge bearing load.

If the steel is calculated as per the formulae given above, it will be observed that the steel for vertical cast process is practically same as that given in Table for vertical cast process, while for the spinning process, for the same dia of pipe, it is much more. Actually, the steel in both the processes should be same, as three edge bearing load is same for spinning and vertically cast process. But, this is not directly possible because thicknesses are different. Hence, in the table enclosed, steel is calculated theoretically, based on present wall thickness. Even then the steel for spinning process is considerably reduced.

Original I.S. 458 was based of Australian Standard for pipe. At that time, Humes were not having theoretical background for design of RCC Pipes. Steel quantities were therefore bases on practical consideration. Indian Specification I.S.458 followed. The quantities by Humes then, now when theoretical approach is available, the quantities of steel should be base on it. Other restriction is wall thickness of pipe. Wall thickness in Indian Standard for Spun Pipe is less as compared to vertical cast pipes, which is based on American Standard where the thicknesses are more. Hence, the first step for Indian Standard IS:458 is to propose the steel as shown in the table for spinning. Changing the wall thickness will be very expensive at this stage as major investment is in moulds. The next step will be to change the thickness of spun pipe, then there will be one table for spun and vibrated pipes.

Presently, it is recommended that the weights of steel for NP2, NP3 and NP4 classes of Spun pipes should be changed immediately which will be equal to the proposed quantities given in the table in Annexure 'A'.

9.7 DESIGN OF CONCRETE PIPE FOR INTERNAL PRESSURE

9.7.1 Typical Example No 3 for P1 Class R.C.C. Pipe

Design Hume Pipe 80 cm dia. to withstand pressure of 20 metre of water. Assume, $n = 6$ and $fc = 20$ kg/cm^2. Use 6 mm dia. hard drawn wire.

Data

I.D of pipe	$D = 80$ cm
Allowance concrete stress	$fc = 20$ kg/cm^2
Allowable stress in hard drawing wire	$fs = 1400$ kg/cm^2
Modular ratio	$n = 6$
Internal pressure	$P = 2$ kg/cm^2
Dia. of hard drawn wire	$S = 6$ mm

Calculation

$$As = \frac{P \times r}{fs} = \frac{2 \times 40}{14000} = 0.0571 \text{ cm}^2/\text{cm length}$$
$$= 5.71 \text{ cm}^2/\text{m length}$$

$$P \times r = fc \, [t + (n - 1) \, As]$$

$$t = \frac{P \times r}{fc} - (n - 1) \, As$$

$$= \frac{2 \times 40}{20} - (6 - 1) \times 0.0571$$

$$= 4 - 0.2855$$

$$= 3.71 \text{ cm}$$

The minimum thickness as per I.S. 458 – 2003 is 4.5 cm.

Area of 6 mm dia wire = 0.2827 cm^2

No of turns/metre $= N = \dfrac{5.71}{0.2827} = 20.2$

Weight of 6 mm dia for 1 metre = 0.2827 × 0.7850 = 0.2214 kg

Weight of Spiral/metre

$$= 20.2 \times 3.1416 \times (0.8 + 0.045) \times 0.2214$$
$$= 11.87 \text{ kg/linear metre}$$

Weight of Spiral as per I.S. 458 – 2003 is 11.94 kg/metre.

Typical Example No 4 for P2 class R.C.C. Pipe

9.7.2 Typical Example No 4 For P2 Class R.C.C. Pipe

Design Hume Pipe 80 cm dia. to withstand pressure of 40 metre of water. Assume, $n = 6$ and $fc = 20$ kg/cm^2. Use 6 mm dia. hard drawn wire.

Data

I.D of Pipe	$D = 80$ cm
Allowance concrete stress	$fc = 20$ kg/cm^2
Allowable stress in hard drawing wire	$fs = 1400$ kg/cm^2
Modular ratio	$n = 6$
Internal pressure	$P = 4$ kg/cm^2
Dia. of hard drawn wire	$S = 6$ mm

Calculation

$$As = \frac{P \times r}{fs} = \frac{4 \times 40}{1400} = 0.1143 \text{ cm}^2/\text{cm length}$$
$$= 11.43 \text{ cm}^2/\text{m length}$$

$$P \times r = fc \, [t + (n-1) \, As]$$

$$t \quad = \frac{P \times r}{fc} - (n-1) \, As$$

$$= \frac{4 \times 40}{20} - (6-1) \times 0.1143$$

$$= 8 - 0.5715$$

$$= 7.4285 \text{ cm}$$

The minimum thickness as per I.S. 458 – 2003 is 8 cm.

Area of 6 mm dia wire $= 0.2827$ cm^2

No of turns/metre $= N = \dfrac{11.43}{0.2827} = 40.43$ say $= 41$

Weight of 6 mm dia for 1 metre $= 0.2827 \times 0.7850 = 0.2214$ kg
Weight of Spiral/metre

$$= 41 \times 3.1416 \times (0.8 + 0.08) \times 0.2214$$
$$= 25.09 \text{ kg/metre}$$

Weight of Spiral as per I.S. 458 – 2003 is 28.54 kg/linear metre.

9.7.3 Typical Example No 5 for P3 Class R.C.C. Pipe

Design Hume Pipe 80 cm dia. to withstand pressure of 60 metre of water. Assume, $n = 6$ and $fc = 20$ kg/cm^2. Use 6 mm dia. hard drawn wire.

Data

I.D of Pipe	$D = 80$ cm
Allowance concrete stress	$fc = 20$ kg/cm^2
Allowable stress in hard drawing wire	$fs = 1400$ kg/cm^2
Modular ratio	$n = 6$
Internal pressure	$P = 6$ kg/cm^2
Dia. of hard drawn wire	$S = 6$ mm

Calculation

$$As = \frac{P \times r}{fs} = \frac{4 \times 40}{1400} = 0.1714 \text{ cm}^2/\text{cm length}$$
$$= 17.14 \text{ cm}^2/\text{m length}$$

$$P \times r = fc \ [t + (n - 1) \ As]$$

$$t \quad = \frac{P \times r}{fc} - (n - 1) \ As$$

$$= \frac{6 \times 40}{20} - (6 - 1) \times 0.1714$$

$$= 12 - 0.857$$

$$= 11.14 \text{ cm}$$

The minimum thickness as per I.S. 458 – 2003 is 12 cm.

Area of 6 mm dia wire = 0.2827 cm^2

No of turns/metre = $N = \dfrac{17.14}{0.2827} = 60.6$ say = 61

Weight of 6 mm dia for 1 metre = 0.2827 × 0.7850 = 0.2214 kg

Weight of Spiral/metre

$$= 61 \times 3.1416 \times (0.8 + 0.12) \times 0.2214$$

$$= 39.03 \text{ kg/linear metre}$$

Weight of Spiral as per I.S. 458 – 2003 is 39.46 kg/metre.

9.7.4 Typical Example No 6 for Design of Pipe for Combined Internal Pressure and External Load (IS 783 – 1985)

Data

Internal dia. pipe	$d = 900$ mm
Wall thickness	$t = 90$ mm

External dia. of pipe $D = 900 + 2 \times 90 = 1080$ mm

Width of trench $B = D + 300 = 1080 + 300 = 1380$ mm = 1.38

Depth of trench of pipe invert = 3 m

Weight of fill material (Wet clay) = $W = 18$ kN/m^3

Bedding and foundation material (see *B*4) = Type B granular earth fill

Load Factor *B*4 page 41 *Ft* = 1.90

Depth of fill material over top of pipe $H = 3 - 1.38 = 1.62$ m

Internal working pressure Pw = 0.20 Mpa

Hydraulic test pressure at factory *Pt* = 0.40 Mpa

Calculation

Vertical load on the pipe due to earth fill on top of pipe $We = Ct\ W\ B^2$

Value of *Ct* is obtained from Fig. 9.4 use the curve and the ratio of H/B
$$= 1.62/1.38 = 1.17$$

Hence for H/B = 1.17 and clay sand D, Ct = 1.0

Calculate earth load on top = $We = Ct \times W \times B^2$ see A1
$$We = 1 \times 18 \times 1.38^2 = 34.28 \text{ kN/m}$$

Load Factor *Ft* = 1.9 (*B*–4)

Load to produce 0.25 mm crack = $We/Ft = 34.28/1.9 = 18.04$ kN/m

For simultaneous internal pressure and external pressure load

$$T = \frac{We}{F} \frac{(Pt)^{13}}{Pt - Pw}$$

$$T = \frac{34.28}{1.90} \frac{(0.4)^{1.3}}{0.4 - 0.2} = 18.14 \times 1.25 = 22.73$$

0.25 crack load when only earth = 18.04 kN/m

0.25 crack load when earth + Water load = 22.73 kN/m

Hence P2 class pipe with NP2 class 0.25 mm pipe load is to be used NP2 class load is 22.73 kN/m

9.8   DESIGN OF CONCRETE CULVERT PIPES

Introduction

Road culvers, despite their apparent simplicity, are complex engineering structures from a hydraulic as well as a structural view point. Their functional adequacy is no better than the estimate of the design flood and the hydraulic design described below must be preceded by a careful flood evaluation together with an assessment of the cost resulting from damage caused by design flood being exceeded.

The hydraulic complexity of culverts is a result of the many parameters influencing their flow pattern. This influence can be summarized by referring to two major types of culvert flow.

9.8.1 Hydraulics of Culvert

The hydraulic evaluation of culvert pipes is based on Manning's formula and the recommended value of Manning's "*n*" for concrete pipe is 0.011.

The flow condition in a culvert is expressed as either inlet control or outlet control. It is essential that the designer investigate both states of flow and adopt the more restrictive flow condition for design.

Inlet control conditions shown in Fig. 9.4 exist in a pipeline where the capacity of the pipeline is limited by the ability of the upstream flow to easily enter the pipeline, a common situation in coastal areas where short culvert lengths on steep grades, are used. The flow under inlet control conditions can be either inlet submerged or un-submerged,

Where culverts are laid on flat grades and empty below the downstream water level, the culvert typically operates with outlet control conditions as shown in Fig. 9.5.

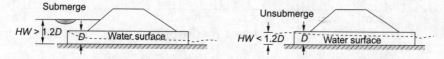

Fig. 9.4

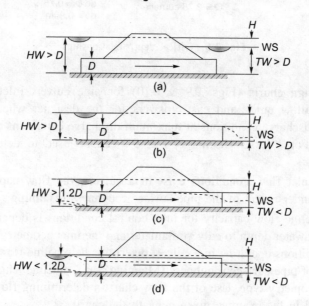

Fig. 9.5 Outlet control

When operating under outlet control conditions, the culvert pipe may flow full or part-full depending on the tailwater depth.

Where the tailwater depth is greater than the pipe diameter, the pipe will typically flow full. Where the tailwater depth is less than the pipe diameter, the design tailwater depth should be taken as the greater of the actual tailwater depth or $(d_c + D)/2$ where d_c is the critical depth for the actual flow discharge, (see Fig. 9.6).

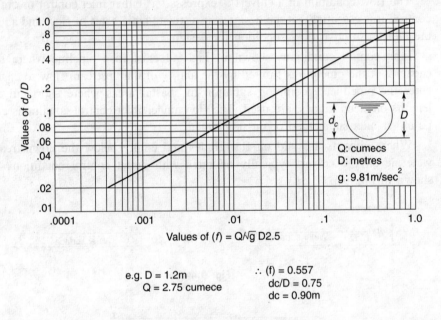

Values of $(f) = Q/\sqrt{g}\, D2.5$

e.g. D = 1.2m
Q = 2.75 cumece

$\therefore$ (f) = 0.557
dc/D = 0.75
dc = 0.90m

Fig. 9.6 Critical depth relationships

The design charts (Figs. 9.9 & 9.10) for pipe culvert inlet and outlet conditions allow quick and easy answers for the designer when evaluating maximum discharge conditions at maximum discharge conditions at maximum headwater. For a lesser discharge, Fig. 9.7 can be used to determine flow characteristics.

Where inlet flow conditions exist in a culvert, the flow capacity of the pipeline is independent of the pipe surface roughness (Manning's 'n').

The pipeline flow capacity for inlet control conditions is dependent on the ratio of headwater depth to culvert diameter and the inlet geometry type. Outlet control conditions operating in a culvert determine he pipeline flow capacity by the effects of pipe surface roughness (Manning's 'n'), pipeline length and slope and inlet geometry type. Use of the flow chart in determining flow conditions is illustrated in the example given beneath the charts.

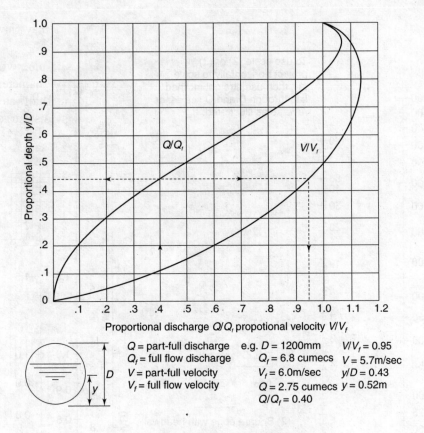

$Q = \text{part-full discharge}$ e.g. $D = 1200\text{mm}$ $V/V_f = 0.95$
$Q_f = \text{full flow discharge}$ $Q_f = 6.8 \text{ cumecs}$ $V = 5.7\text{m/sec}$
$V = \text{part-full velocity}$ $V_f = 6.0\text{m/sec}$ $y/D = 0.43$
$V_f = \text{full flow velocity}$ $Q = 2.75 \text{ cumecs}$ $y = 0.52\text{m}$
   $Q/Q_f = 0.40$

Fig. 9.7 Flow relationships

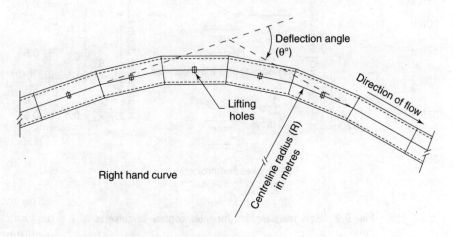

Fig. 9.8 Flush joint splays in curved pipeline alignment

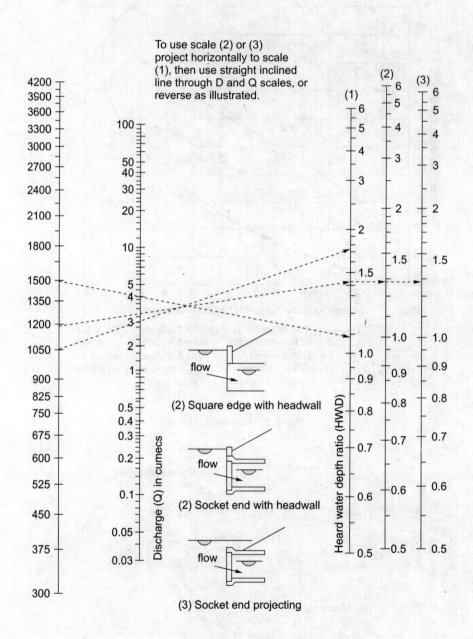

Fig. 9.9   Flow relationships for inlet control in culverts

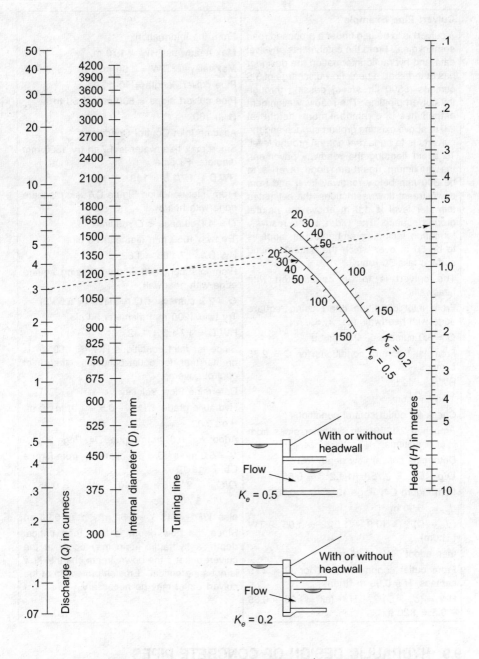

Fig. 9.10 Flow relationships for outlet control in culverts

Culvert Pipe Example

A culvert is to be laid under a proposed road embankment. From the catchments physical data and hydraulic information the designer has determined a peak flow discharge of 5.5 cumecs (5500 litres/sec) passing through the culvert pipeline. The roadway alignment establishes the embankment height at 2.0 m above existing ground surface and the culverts is to be laid at natural ground level. To avoid flooding the roadway pavement, the maximum upstream flood level is to be 300 mm below roadway level and from downstream flow restrictions, the estimated tailwater level is 1.0 m above the natural ground surface. The width of the roadway formation including embankment slope is to be 50 m over which the natural ground surface falls 500 mm.

The culvert is to be constructed with headwalls.

From inlet condition chart using 'square edge with headwall' (i.e. k_e = 0.5);

Q = 3.0 cumecs > Q required

From inlet control conditions for Q = 2.75 cumecs

HW/D = 12.5

∴ HW = 1.5m

Check for outlet control conditions

Proposed 2/1200 mm diameter pipes from inlet conditions

Determine critical flow depth (d_c)

$Q/g \ D^{2.5}$ = 2.75/9.81/1.2$^{2.5}$ = 0.557

from Figure C4, Page 12 d_c/D = 0.75

d_c = 0.90 m

$(d_c + D)/2$ = (0.9 + 1.2)/2 = 1.05 > TW (=1.0m)

then adopt TW = 1.05 m

From outlet condition chart, for Q = 2.75 cumecs, H = 0.65 m then

HW = $(d_c + D)/2 + H$ – fall = 1.05 + 0.65 – 0.5 = 1.20 m

From the information,

Max headwater HW = 1.70 m,

Max tailwater TW = 1.00 m,

Pipe culvert length = 50 m,

Pipe culvert slope = 500 mm in 50 m (1 in 100)

Assume Inlet Control Conditions

Since max headwater is 1.7 m try 1500 mm diameter FJ pipe

HW.D = 1.7/1.5 = 1.13

From inlet condition Figure C7 using 'square edge with headwall',

Q = 4.1 cumecs < Q required.

Try twin 1050 mm diameter FJ pipe

HW/D = 1.7/1.05 = 1.62

From inlet condition Figure C7 using 'square edge with headwall',

Q = 2.2 cumecs < Q required (= 5.50/2)

Try twin 1200 mm diameter FJ pipe

HW/D = 1.7/1.2 = 1.42

Since for inlet conditions HW (= 1.50 m) is greater than for outlet conditions, then inlet control governs.

Determine Flow Velocity

Hydraulic grade = (1.5 + 0.5 – 1.0) in 50 m i.e., 0.02

Adopt K_s = 0.6, from Figure D4, Page 20

V_f = 6.0 m/se, Q_f = 6.8 cumecs from Figure C5, Page 12

Q/Q_f = 2.75/6.8 = 0.40 V/V_f = 0.95

∴ V = 5.7 m/sec

also Y/D – 0.43 Y = 0.52 m < d_c (= 0.90 m)

Since the flow depth is less than critical depth, a hydraulic jump may occur at the culvert outlet if the downstream channel flow is not supercritical. Erosion protection at the culvert outlet may be necessary.

9.9 HYDRAULIC DESIGN OF CONCRETE PIPES AND PIPELINES

9.9.1 Discharge through Concrete Pipelines

When the length of a concrete pipeline exceeds Approximately 100 times its diameter and the pipe is Flowing full, the discharge can be calculated by use of the Hazen-Williams formula. This should be Applied after making due allowance for head losses Other than the friction head.

$$* \; V = 1.318 \; C/1.516 \; R^{0.65} \; H^{0.54}$$

Where,

V = Mean velocity in meter per second.

H = Friction head in meter per meter length = hydraulic slope.

R = Hydraulic radius, meter

  = $D/4$ for pipe running full

    (D = internal diameter of pipe in meters)

C = Hazen-Williams coefficient

The value for C for spun concrete pipes of smooth bore is 140. A comparison of coefficients used in other formulae is as follows:

Manning n	Hazen-Williams C	Scobey C_s
0.013 to 0.014	95 to 125	0.32
0.011 to 0.012	110 to 140	0.36
0.010 to 0.011	120 to 150	0.40

The diameters considered on the graph are in Multiples of 25 mm. when bore diameter are in Imperial dimensions (multiples of 25.4 mm) the Discharge under given circumstances is 4.3% higher than Indicated by the graph.

9.9.2 Pipeline Design

Pipeline is subjected mainly to

  (i) Internal Hydraulic Pressure

 (ii) External Loads

These are discussed below:

Hydraulic Design of Pipeline

This chapter provides formulas and guidelines to aid in the hydraulic design of concrete pressure pipe. The three formulas typically used to determine pipeline capacity are presented and compared. Factors that contribute to the decrease with age of pipeline carrying capacity are also described. The methods used to determine minor and total head losses are presented, as well as the methodology for determining an economical pipe size. Throughout this chapter, discussions will be limited to combinations of pipe sizes and flows that provide a range of velocities and diameters most commonly used with concrete pressure pipe.

9.9.3 Flow Formulas

The formulas commonly used to determine the capacity of pipelines are the Hazen-Williams formula, the Darcy-Weisbach formula, and the Manning formula. Many other flow formulas may be found in technical literature, but these three are most frequently used. It is impossible to say that any one of these formulas is superior to the others for all pipe under all circumstances, and cannot be expected identical answers to a problem from all these formulas. Judgment must be used when selecting a flow formula and the roughness coefficient for a particular hydraulics problem pipe.

(A) The Hazen-Williams Formula

The empirical Hazen-Williams formula is probably the most commonly used flow formula in the waterworks industry. The basic form of the equation is:

in PS units	in Metric units	
$V = 1.318 \, CR^{0.63} \, S^{0.54}$	$V = 0.849 \, C \, R^{0.63} \, S^{0.54}$	(9.1)

Where,

V = Velocity, in feet per second

C = Hazen-Williams roughness coefficient

R = Hydraulic radius, in feet, which is the cross-sectional area of the pipe divided by the Wetted perimeter, i.e., for circular pipe flowing full, the internal diameter in feet divided by 4.

S = Slope of the hydraulic grade line, in feet per foot calculated as h_L/L, where h_L equals head loss in feet occurring in a pipe of length L in feet.

For circular conduits flowing full, the equation becomes

$$V = 0.550 \, C \, d^{0.63} \, (h_L/L)^{0.54} \qquad (9.2)$$

Where,

V = Velocity, in feet per second

C_h = Hazen-Williams roughness coefficient

d = Inside diameter of the pipe, in feet

h_L = Head loss, in feet

L = Pipe length, in feet

A detailed investigation of the available flow test data for concrete pipe was preformed by Swanson and Reed (1963), who concluded that the Hazen-Williams formula most closely matches the results for the range of velocities normally encountered in water transmission.

Table 9.1 Hazen and Williams formula in different form

In terms of	Velocity, fps	Flow rate, cfs
General equation	$V = 0.550\,C_h d^{0.63}\,(h_L/L)^{0.54}$	$Q = 0.432\,C_h d^{2.63}\,(h_L/L)^{0.54}$
Head loss, ft	$h_L = 3.021(L/d^{1.167})(V/C_h)^{1.852}$	$h_L = 4.726\,(L/d^{4.870})(Q/C_h)^{1.852}$
Pipe diameter, f	$d = 2.580(V/C_h)^{1.587}(L/h_L)^{0.857}$	$d = 1.376(Q/C_h)^{0.380}(L/h_L)^{0.205}$
Pipe length, ft	$L = 0.331\,h_L\,d^{1.167}(Ch/V)^{1.852}$	$L = 0.212\,h_L\,d^{4.870}(Ch/Q)^{1.852}$
Roughness coefficient	$C_h = 1.817(V/d)^{0.63}(L/h_L)^{0.54}$	$C_h = 2.313(Q/d)^{2.63}(L/h_L)^{0.54}$

A statistical analysis of Swanson and reed's test data led to the development of the following equation for determining C_h:

$$C_h = 139.3 + 2.028d \tag{9.3}$$

Where,

C_h = Hazen-Williams roughness coefficient

d = Inside diameter of pipe, in feet.

Equation 3.3 can be used to calculate a C_h value for any size pipe; however, for design purpose the following conservative values are suggested:

Table 9.2 Recommended value of C_h in Hazen and Williams formula

Diameter, in	C_h . Value
16 to 48	140
54 to 108	145
114 and larger	150

These values are applicable to concrete pipelines in which the fitting losses are a minor part of the total loss and for pipelines free from deposits or organic growths that can materially affect the pipe's carrying capacity.

(B) The Darcy-Weisbach Formula

An alternative to the Hazen-Williams's formula is the Darcy-Weisbach formula, which is

$$h_L = f\,LV^2/d^2 g$$

Where,

h_L = Head loss, in feet of water

f = Darcy friction factor

L = Pipeline length, in feet

V = Velocity, in feet per second

d = Inside diameter of the pipe, in feet

g = Gravitational constant, 32.2 ft/s^2

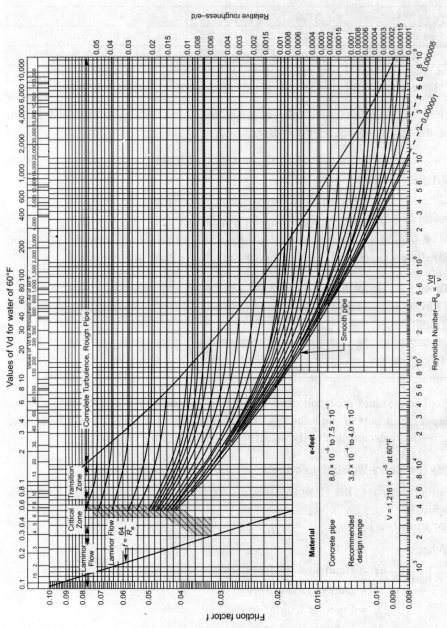

Fig. 9.11 The moddy diagram for friction in pipe

The Darcy friction factor (f) can be determined from the Moody diagram, presented in Fig. 9.11. To use this diagram, the Reynolds number R_e and the relative roughness (e/d) must first be calculated. Both terms are defined on pre page.

The Reynolds number is a function of the flow in the pipe and may be calculated as

$$R_e = dV/v$$

Where,

> R_e = Reynolds number
>
> d = Inside diameter of the pipe, in feet
>
> V = Velocity, in feet per second
>
> v = Kinematic viscosity of the fluid, in square feet per second.

(All these are taken from originals, hence are in British units)

The kinematic viscosity of water of water at various temperatures from freezing to boiling is presented in Table 9.2.

The relative roughness e/d of a pipe is a function of the absolute roughness (e) of the interior surface of the pipe and the pipe diameter (d) Values of the absolute roughness (e) for concrete pipe range from 8.0×10^{-5} to 7.5×10^{-4}, and the recommended design range is 3.5×10^{-4} to 4.0×10^{-4}. Once the Reynolds number and the relative roughness are determined, the moody diagram (Fig. 9.11) can be used to determine values of the Darcy friction factor (f), which may then be used to solve the Darcy–Weisbach formula.

If an analysis solution for the Darcy friction factor (f) is preferred, it may be obtained by iteration from the Colebrook–White equation.

$$\frac{1}{\sqrt{f}} = -2 \log_{10} [e/3.7d + 2.51/R_e\sqrt{f}]$$

Where,

> f = Darcy friction factor
>
> e = Absolute roughness of the pipe, in feet
>
> d = Inside diameter of the pipe, in feet
>
> R_e = Reynolds number

or in metric units

$$f = \frac{0.25}{[\text{Log } e/3.7d + 5.7/R_e^{0.9}]^2}$$

Where,

> f = Darcy friction factor
>
> e = Absolute roughness of the pipe, in m
>
> d = Inside diameter of the pipe, in m
>
> R_e = Reynolds number
>
> e/d = Relative roughness

Measurement of absolute roughness

It is now possible to measure absolute roughness of concrete surface at site by using portable "profilometer". Some observations taken on cement mortar lining of 1400 mm diameter steel pipe are in Table 9.3.

Table 9.3 Kinematic viscosity of water at different temperatures

Temperature °F	Kinematic Viscosity (v) ft²/s	Temperature °C	Kinematic Viscosity (v) m²/s
32	1.931×10^{-5}	0	1.79×10^{-6}
50	1.410×10^{-5}	10	1.31×10^{-6}
68	1.09×10^{-5}	20	1.01×10^{-6}
86	9.57×10^{-6}	30	0.89×10^{-6}

9.10 VENTILATION OF CONCRETE PIPELINE

9.10.1 Necessicity of Air Release Valves on Pipelines

In water pipeline it is utmost important to provide adequate means of releasing air from it and introducing air into it. The presence of free air in a pipeline in services causes many troubles, principal among them is reducing the discharging capacity of that line. There are many instances of pipe lines which have become fully effective; only after a careful study of ventilation requirements. Successful evacuation of air was achieved either by addition of air valves or increasing the capacity of air valves.

Study of "C" values in Hazen and William's formula achieved in overseas pipeline, based on actual observation on pipelines in service, indicates that the values of "C" achieved in India are on lower side. Actual values of "C" achieved in India for same pipelines, against overseas pipe lines are indicated below.

Table 9.4 Value of C_h value as used in India and overseas

Dia.	Value of "C_h"		Less as compared to overseas percent	Year	Pipeline Number	Type of pipes
	*Overseas	India				
600	145	120	17.24	1979	1	Concrete
900	150	123	18.00	1967	2	Concrete
1200	150	127	15.60	1975	3	Concrete
1676	156	130	16.60	1978	4	Concrete
2000	156	135	13.46	2000	5	Steel with c.m. lining
2200	156	139	10.90	2002	6	Steel with c.m. lining

*"C" values are from Swanson and Reed (1963), Manual M-9 AWWA.

9.10.2 Air in Water and its Effect

It is common belief amongst engineers that mere placement of air valves, at peak points on pipeline is sufficient, to get rid of air. Research over the past 35 years, however, has proven this method of approach to be inadequate for total air management This is due to unpredictable nature of air and inherent shortcomings, in conventional air valves design; which does not solve the problem of air retention; but can aggravate the phenomenon related to its presence.

A typical pipeline will contain dissolved air and entrained air: the dissolved air can amount to 2% of system volume, while the entrained air can vary up to 10% of the flow, depending on the degree of turbulence at the system inlet and ingress through joint defects. The solubility characteristics of air in water are very relevant, since a fall in pressure or rise in temperature reduces water capacity to retain dissolved air.

In a typical pipeline system, air is constantly being released due to pressure reduction following changes in pipeline level and restriction in the line such as partly closed isolating valves, pressure reducing valves orifice plate or similar obstructive devices.

Once in the system, air tends to collect at high points although there are conditions, which will allow air bubble to remain static in the pipeline. However, if the fluid velocity is in excess of 2 m/sec., it is probable that the flow will sweep the air at a ventilation point.

Discharge is adversely affected as can be seen from the test conducted by Council of Scientific and Ind. Research - South Africa that flow capabilities of 4 different 80 mm diameter air valve designs with different large orifice diameters at equivalent differential pressures.

Table 9.5 Effect of large orifice on air valve performance

Country of Manufacture	UK	Israel	Turkey	South Africa
Valve nominal diameter	DN 80	DN 80	DN 80	DN 80
Dia. of large orifice (mm)	65	32	65	80
Max. discharge at 0.05 bar	83	36	81	156

Discharge through 80 mm dia orifice is 61% higher than 65 mm diameter. Hence, the actual dia of large orifice valve is very critical.

9.10.3 Types of Air Release Valves

Air is entrained in water by many ways; by vortices in the pump suction reservoirs or merely by absorption at exposed surface at air pockets or in the suction reservoir. The air may be released when there is a drop in pressure, either along a rising main or when the velocity is increased through a restriction, such as partially closed valves. An increase in temperature will also cause air to be released from solution. Small orifice air release valves are installed on the pipeline to bleed off the air, which comes out of solution.

Small orifice air valves are designated by their outlet connection size usually 12 mm to 50 mm dia. This has nothing to do with the air release orifice size which may be from 1 to 10 mm diameter. The larger the pressure in the pipeline, the smaller need to be orifice size. The volume of air to be released will be a function of the air contained which is on the average 2% of the volume of water (at atmospheric pressure).

The small orifice air release valves are sealed by a floating ball, when certain amount of air has accumulated in the connection on top of pipe, the ball will drop and release the air. Small orifice valves are often combined with large orifice air vent valves on a common connection on the top of the pipe. The arrangement is called double air valve. An isolating sluice valve is normally fitted between the pipe and the air valves.

Double air valves (Fig. 9.12) should be installed at peaks in the pipeline, both with respect to the horizontal and max. hydraulic gradient. They should also be installed at the ends and intermediate points along length of pipeline which is parallel to the hydraulic gradient line. It should be borne in mind that air may be dragged along in the direction of flow in the pipeline, so it may even accumulate in sections falling slowly in relation to the hydraulic gradient. Double air valve should be fitted every 1/2 km to 1 km along descending sections, preferably on manholes, especially at points where the pipe lines dips steeply.

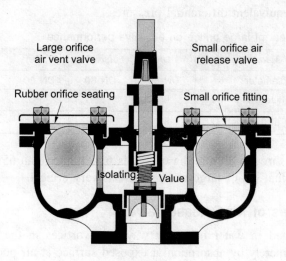

Fig. 9.12 Double air valve

Air valves should also be installed along ascending lengths of pipe lines where air is likely to be released from solutions due to lowering of the pressure, again especially at points of decrease in gradients. Other places where air valves are required are on the discharge side of pumps and at high points and up stream of orifice plates and reducing tapers.

9.10.4 Conventional Air Valve Designs and their Limitations

Conventional air valves fall in one of two categories namely,

(a) Non kinetics air valves

These are usually air valves with spherical floats, which have tendency of closing prematurely and often retain large pockets of air in the pipeline. This phenomenon is known as Dynamic closure.

(b) Kinetic air valves

These were developed primarily to overcome the premature closure phenomenon that plagues conventional Non Kinetic Design. This is achieved by altering the internal configuration of the valve and therefore its aerodynamic characterization in such that, during air discharge the large orifice float is biased towards the inlet orifice thereby preventing premature "Dynamic closure" of the outlet orifice. Hydrodynamic principle is used to control the ball. The risk of closing the valve when the air is released is completely eliminated by the provision of nozzle. This is an improved version of large Orifice air valve.

The valve is so designed that it remains open even when the pressure falls sufficiently and water recedes from the valve. The valve drops into a nozzle which is so arranged that as to cause the escaping air stream to flow around it in such a manner that the resultant pressure on the ball is downwards and increase as the emergent air velocity increases. The pressure distributions caused by the stream line flow of air through the nozzle all around the ball thus gives a positive downward effort on the ball so that air at any velocity can be safely discharged without risk of ball being caught up in the air steam.

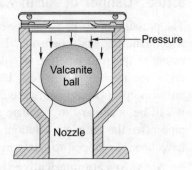

Kinetic air valve

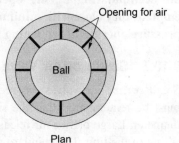

Plan

Fig. 9.13   Kinetic air valve

Problems created due to discharge of air at high velocity.

(i) Air valve discharging air at high differential pressure and velocities, causes high damaging transient pressure. This is due to water flow entering the valve suddenly being arrested by the large orifice control float seating on the large orifice. The effect on the pipeline dynamics is equivalent to the rapid closure of an isolating valve. It is recommended therefore that the discharge differential pressure across the large orifices of Kinetic air valve to be 0.5 N/mm^2 differential pressure in order to prevent damaging high pressure transient being occurring.

(ii) Limitations of the large orifice - Many Kinetic air valve designs are mere modifications of Non Kinetic valve using spherical floats to seal off the large orifice.

9.10.5 Air Valve Performance Capabilities

It has been accepted standard to specify an Air valve in terms of nominal inlet diameter only. This is peculiar practice considering that the discharge and intake performance of any particular air valve is dependent on design and internal configuration, the size of large orifice in relation to nominal size of valve. The slope and mass of large orifice control float and the maximum allowable differential pressure across large orifice. Performance may vary diametrically between different manufacturers with the same nominal inlet diameter.

9.10.6 Danger of Air in Water

The pressure of air in a pipeline can cause delays in filing, reduction in capacity, water hammer and corrosion. If maximum efficiency is to be attained, it is essential that the pipeline be filed quickly with all entrained air removed to allow the system to run full. Unless the system is an ideal profile, having a uniform upward gradient in the direction of flow and is free of all obstructions, it will be necessary to ventilate the pipeline to achieve this result. While vent pipe offer the simplest solution, these are almost impracticable and air-valves have been developed to perform this function.

Air valves automatically exhaust air from pipeline during filling, release air accumulating during operation and admit air when the pipeline is empted; they close with minimum of fluid loss and remain drop tight during pressure operation, reopening whenever a depression occurs in the pipeline.

9.10.7 Sizing of Air Valve

Small orifice air valves are required to discharge air at pipeline pressure. It must be remembered that the size of the valve, is no indication of its orifice diameter. Large orifice air release valve is required to vent air during the initial filling of the line. The standard approach to sizing has been to provide adequate valve diameter to vent air at volumetric rate equal to pumping rate by converting the pressure rate to an equivalent cubic meter per minute. A common design specification is for a cluster of at least two valves, but not more than four. One of the valves then serves as a spare with the discharge capacity divided among the other. It is important to emphasize again the tendency of large orifice unit to blow shut. To alleviate this problem, valve diameter must be large enough to limit the pipeline air pressures to vary low values at the order of 1-2 psi. Large orifice valve require very gradual filing rates to minimize the water hammer that result when air is closed. It is important to provide chambers to all the air valves properly vented and drained. An improperly vented chamber may become pressurized during the discharge of air, and during the period of inflow the chamber may be subjected to severely negative pressure. The size of valves tends to be empirical. Reference should be made to manufacturer's literature. Where the actual air flow requirements are known the valve size can be calculated from the formula from flow up to differential pressure of 0.9 bar. Refer illustrative example for the calculation of size of valve.

9.10.8   Indispensability of Air Valve

Water is a very precious commodity and huge investment is involved to transport and supply it to the large population of this country. A careful evaluation of the characteristics and function of air release valves and their applications can lead to an investment that will bring immediate return without major cost consideration. It is the duty of Engineer to see that the discharging capacity of our pipelines should be of international standards.

Even if the air valves available in the country are properly located; 5-7% increase in discharging capacity of existing pipelines is possible. If further improvement is desired - which is a must, conventional air valves (whose design is 100 years old) have to be replaced by improved design valves, common in overseas countries which are now available in India.

9.10.9 Illustrative Example – Calculation for Size of Large Orifice Valve

Basically there is no precise procedure to calculate the discharging capacity i.e., the size of valve, as the conditions in pipe vary frequently.

An indicative procedure to calculate the size of valve is given below. This is based on the information available in the old catalogues of Glenfield and Kennedy. Graphs are also provided to select the size of valve. The units used in that literature are British units hence the same are used in this example.

(a) Air discharging capacity of valve

$$Q = CdD^2 \sqrt{P}$$

Where,

Q = Flow of air through valve cub.ft/min

Cd = Valve constant (about 20)

D = Dia. of valve in inch

P = Pressure difference across orifice

(Recommended pressure differential across orifice is 10lb/in^2)

For 6" valve

Flow of Air $Q = 20 \times 36 \sqrt{10} = 2275$ cft/min

When pressure differential is $10lb/\text{in}^2$ (0.07N/mm^2), inside pressure will be $14.7-10 = 4.7\text{lb/in}^2$.

Also from the graph, the discharge for inside pressure of 4.7 lb/in^2 is 2300 cft- this agrees with the calculated figure of 2275 cft/min

(b) Diameter of air valve

To find out diameter of the valve

Given,

- Pipe dia. – 4 ft
- Spacing between valves –1500 ft
- Velocity – 5 ft/sec

Volume of water between two valves

$$= 0.7854 \times (4)^2 \times 1500 = 18850 \text{ cft}$$

Time in minute to travel 1500 ft $= \dfrac{1500}{5 \times 60} = 5$ mins

Discharge of valve required $= \dfrac{18859}{5} = 3770$ cft/min

Diameter of valve D

$$D^2 = \frac{Q}{Cd\sqrt{P}} = \frac{3770}{20\sqrt{10}} = 59.6$$

D = 7.72" i.e., 8" dia value will be required.

(c) If the velocity is reduced to 3 ft/sec

Time required to travel 1500 ft will be $\dfrac{1500}{3 \times 60} = 8.33$ mins

Discharge of valve required will be $\dfrac{18850}{8.33} = 2263$ cft/min

Dia. of valve required will be

$$D^2 = \frac{2263}{63.24} = 35.78$$

D = 5.98" i.e., 6" dia value will be required.

(d) If the spacing is reduced to 1200 ft and velocity considered 3 ft/sec

Time required to travel 1200 ft will be $\dfrac{1200}{3 \times 60}$ = 6.666 mins

Volume of water between two valves = $0.7854 \times (4)^2 \times 1200 = 15080$ cft

Discharge of valve required will be $\dfrac{15080}{6.666}$ = 2263 cft

$$D^2 = \frac{2263}{63.24} = 35.80$$

D = 5.98" i.e., 6" dia valve will be required.

It is clear, that capacity of Air valve depends upon the velocity of water and the spacing between valves; Higher velocity require higher capacity. It also dispends upon the constant "C" for the valve; which must be obtained from the manufacturer.

When pumping is started all the valves will be operative for discharging air, but when water reaches near the end; only one valve is operative and the above calculations are for this condition.

Presently the valves are located at 500 meters (1640 ft) in that case the capacity will be slightly increase, depending upon the velocity.

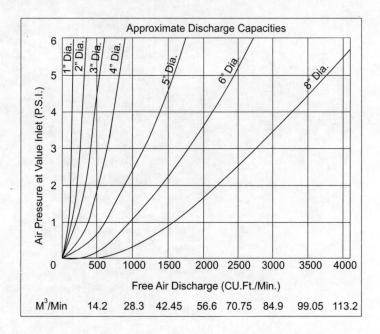

Fig. 9.14

CHAPTER 10

Laying and Field Testing of Pipes with Rubber Ring Joints

10.1 PRELIMINARY WORK

Longitudinal section of the pipe line should be studied carefully and the following points should be checked:

(i) The maximum static head in the case of gravity main is not more than 2/3 of the test pressure. For pumping line, maximum head at any point in a line should not be more than half the test pressure. If in any zone the pressure is found to be more than above, necessary action need to be taken.

(ii) It should be seen whether necessary provisions for counteracting effect of water hammer are provided. If they are not, action is called for.

(iii) It should be seen that, air valves are provided at peak points for the releasing air to prevent air lock. There should be scour valve for scouring the line, near about the nallahs. Distance between two air valves should generally be not more than 1000 mtrs.

(iv) It should be seen whether there are sluice valves provided about 2 to 3 kilometers apart for detaching line from the main line for repairs etc.

(v) The pipeline should preferably be horizontal for a distance of about 15 m on both sides of the air valve for easy fixing of the air valve.

(vi) A curve of 1/2° in one pipe of 5 m length can be negotiated, but continuous curvature should be avoided. The curve can be given for 10 to 15 pipes and not more than that at a time.

(vii) Whenever the pipe line is to cross nallah and it is to be taken under the nallah bed, pipe should be suitably anchored to concrete foundation to prevent uplifting of the pipe line during flood. If it is carried on pillars it should be seen that they are on firm foundation and pipeline to above maximum flood level. The concreting to be done after successful line testing.

(viii) As far as possible pipe should not be exposed to alternate hot dry atmospheric conditions. There should be at least a cover of 1 m of earth. Rubber rings are deteriorated when exposed to sun light.

A programme should be then drawn for laying of pipes. For this, points where gaps should be left for testing should be decided. The laying should be started from these gaps.

It is economical to leave gaps at air valve tees or sluice valves, and not at scour valves or at bends. These gaps will have to be used for testing lines on sections. Sections of 800 to 1600 m are generally suitable. While deciding the gaps, availability of water nearby should also be ascertained. The starting point of laying should be from down hills towards up-hill. It is thus desirable to start laying from scour valve and to proceed on both sides towards air valve keeping sockets up-hill.

After the planning for laying is done, written instructions should be sent to the factory, indicating location (i.e. chainage) giving number of pipes to be carted with necessary heads. This is very important, otherwise the pipes may be carted wrongly and may required reshifting. Instructions should also be given on which side of the trench the pipes should be unloaded. This will avoid reshifting of pipes at the time of laying.

10.2 EXCAVATION

For proper excavation, the following points should be observed:

  (i) Bench marks should be established at every 800 to 1000 m along the alignment.
 (ii) Ground level should be checked by the Engineer In-charge to ascertain whether L-section is correct.
(iii) Formation levels given by client should be checked.
 (iv) Width of the trench should be decided and marked at site.
  (v) Before commencing the excavation, the site on which earth is to be dumped should be fixed up. The pipe should be unloaded on one side of the trench and the earth preferably to be dumped on the other side. This will enable the lowering of the pipe easily.

10.3 LAYING OF PIPES

10.3.1 Laying of Pipes by Machine

Concrete pipes are laid in the same manner as pipes composed of other materials; however, the properties of this pipe material and its rubber ring jointing requirements involve a number of special features with regard to the laying procedure.

(a) The higher wall thickness of the pipe relative to the diameter and the normally unnecessary protective coating permit the use of bedding material in which stones up to a grain size of 50 mm are not detrimental to the satisfactory bedding of the pipe. The influence of the resulting point loads is negligible in view of the existing wall thickness.

(b) Surrounding the pipe by a special sand bed is for the same reason not required. The excavated material can normally be re-used immediately.

(c) The finished pipe joint can be reliably tested immediately by examining the position of the roll-type rubber ring by inserting feeler gauge, thus permitting the trench to be backfilled directly following laying. In this manner, lengthy trench maintenance, with the possibility of bottom deteriorating, is avoided.

(d) The hinge-type jointing of the pipes provides a number of advantages for the pipeline. It must not, however, be impaired by construction measures.

When connecting the pipe to concrete structures such as manholes, casings, abutments and the like, the joint and thus the hinge must be located immediately outside.

10.3.2 Pipe Handling

Pipes are handled by means of cables, straps, clamps or grabs. If the pipe is provided with outer insulation, cables may not be used during handling. Pipes may not be rolled or dropped from the transport vehicles. Any and all damage to the pipe during unloading and storing must be avoided. Since damaged pipe ends can result in pipe joint leakage.

Damaged pipes may not be included in a pipeline until reusability has been clarified with the pipe supplier.

Fig. 10.1 Pipe handling

10.3.3 Handling and Care of Rubber Rings

The correct procedures for the proper storage, maintenance, and cleaning of rubber rings requires, among other things, that rubber rings be stored in a cool, dark, dry and dust free environment. The room temperature should not exceed +20°C or drop below –10°C. The storage location may not be exposed to direct sunlight or radiator appliances. In winter, the rubber rings must be allowed to warm up in a room heated to maximum 20°C until just before fitting in order to provide for the necessary flexibility.

The rubber rings may not come into contact with fuels or lubricants and must be stored free of tension i.e., not subject to severe deformation.

10.3.4 Checking and Preparing the Pipe Ends, for Mounting the Rubber Ring

The pipe ends, especially the sealing surfaces of the socket, must be carefully cleaned on the outside and inspected for any damage. Minor chipping can be corrected on site the laying technician of the pipe supplier or according to his instructions. Seriously damaged pipes, on the other hand, may not be installed unless they have been explicitly cleared for use.

By using the roll-type rubber ring, the rubber ring must be mounted on the groove of the spigot. The confined rubber ring is placed directly in the groove. Both rubber ring and the inside of the socket are coated with a lubricant (details concerning make or type can be obtained from the pipe supplier).

The rubber ring should not be twisted during mounting, no matter what type is being used. Just as important is the uniform tension of the rubber ring along the entire circumference of the pipe. Prior to fitting, the rubber rings should be inspected for dirt and/or damage such as buckling, scouring, or cracking. Damaged rubber rings must never be used.

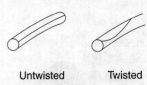

Untwisted   Twisted

Fig. 10.2 Rubber ring twisted and untwisted

10.3.5 Finishing the Trenching Bottom

The bottom of the trench must possess a prerequisite load-bearing capacity to ensure that no setting will occur after pipe laying and during pipeline operating which could result in damage to pipeline. Where the ground is a poor load-bearer, suitable measures must be taken to increase the load-bearing capacity, e.g., soil stabilization.

In rocky or stony ground, the bottom of the trench be excavated at least 15 cm deeper and that this excavation be replaced by a stone less layer. In order to avoid uneven compression loads in the bedding, an adequate stone less soil is particularly important in the area of the pipeline joints.

Socket pits of 15 to 30 cm depth, depending on the pipe diameter, are to be provided for the socket of the pipe so that, during placement of the pipe, the socket does not come in contact with bedding.

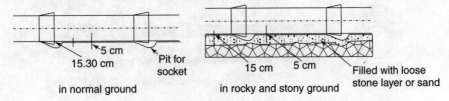

Fig. 10.3 Pipe bedding

10.4 LAYING PROCEDURE, PIPE JOINTING, DEVIATIONS

Experience has shown that the sequence of steps described in the following illustrations comprise the most expedient and economical procedure for the laying of prestressed concrete pipes.

In Phase 1, the trench is prepared by leveling and cleaning the bottom and excavating the socket pit for next pipe (Phase-1).

Suspended from a crane, the pipe is oriented in the direction of the pipeline and so placed into the previously laid pipe, that the rubber ring coincides uniformly with the circumference of the beveled guide of the socket, i.e., so that the ends of the two pipes are centered with respect to one another. It is basically of no importance, as regards spigot or socket end, in which direction pipe laying is carried out. However, it is more common – because more simple – to insert the spigot of the pipe to be laid into the socket of the pipe already laid (Phase-2).

Pulling the pipe home can be accomplished by means of suitable pulling/hoisting devices, in which the pipe is suspended in the device to eliminate friction between the pipe and the bedding (Phase-3).

It may prove operationally advantageous, especially for pipes having a diameter of 800 mm and greater, to guide the cable of the pulling/hoisting device into the pipe or inside of the pipeline.

60% of the weight of the pipe can be assumed as a rough guide for the force required to pull the pipe home.

Roll-on type rubber ring

At the start of pulling the pipe home, due care must be taken to ensure that the rubber ring rolls uniformly onto the entire circumference of the pipe in

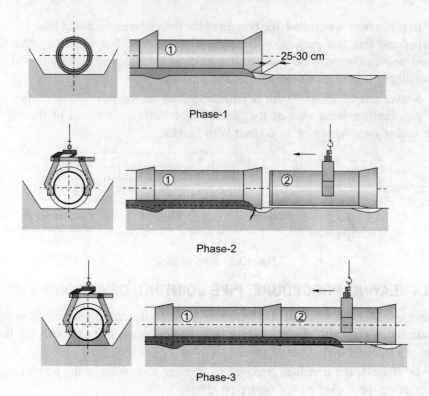

Phase-1

Phase-2

Phase-3

Fig. 10.4 Phases during laying of pipes

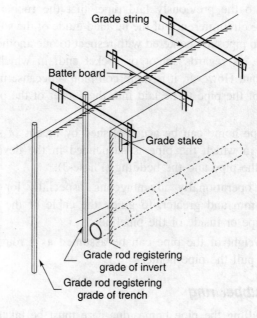

Fig. 10.5 Batter board set up

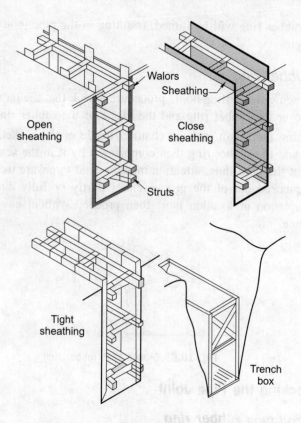

Fig. 10.6   Types of sheathing and shoring

particular at the bottom of the pipe. When pulling home, the pipe being installed must be moved in the centerline of the pipe already laid.

Should the rubber ring be rolled into place unevenly or should it slip, the tension necessary for pulling home may greatly increase. In this case, the pulling home procedure must be interrupted and, after repositioning the rubber ring, reattempted. If the pipes are pulled fully together with the rubber ring compressed in between, there is danger of the socket fracturing; above all,

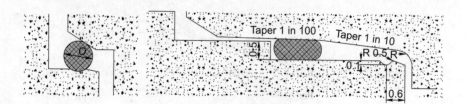

Fig. 10.7   Roll-on type rubber ring

however, the rubber ring will be ruined, resulting in the pipe joint being leaky from the outset.

Confined rubber ring

When commencing the pulling-home procedure, check that the lubricant is still fully present over the rubber ring and the inside of the rubber ring.

   When placing the spigot into the chamfer inside of the socket, the pipe is first centered and the rubber ring then compressed by it in the seating groove. In this phase of the procedure, attention must be paid to ensure that the rubber ring is not squeezed out of the groove either partly or fully due to uneven pressure. Completion of location must then proceed without any appreciable increase in force.

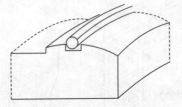

Fig. 10.8   Groove seat rubber ring

10.4.1   Checking the Pipe Joint

Joint with roll-type rubber ring

After the pipe joint has been established, a feeler is used to check the proper location of the rubber ring; this is done from the outside through the socket gap. The feeler is a strip of sheet metal roughly 15 mm wide and maximum 0.5 mm thick; alternatively, a length of wire of a sheet gauge can be employed. This simple tool is sufficient for checking around the entire circumference of the pipe, whereby particular attention must be paid to the bottom area of the pipe. The proper position of the rubber ring is shown by value "b" in the table. If the rubber ring is not in the proper position, the pipes must be pulled apart and the pulling-home procedure repeated.

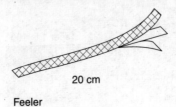

20 cm

Feeler

Fig. 10.9   Feeler gauge for checking depth of rubber ring

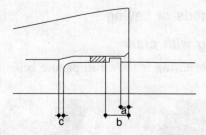

Fig. 10.10 Socket with rubber ring position of laid pipe

Typical socket check report

Project designation:...........................

DN:.........................

Date:........................

Name:...........................

Station:...........................

From..........................   to ....

Ultimate data for a, b, c

Fitting-pipe-number	Pipe joint					
	Concrete ring "a"		Rubber gasket "b"		Gap "c"	
	max.	min.	max.	min.	max.	min.
	⊕	⊕	⊕	⊕	⊕	⊕
	⊕	⊕	⊕	⊕	⊕	⊕
	⊕	⊕	⊕	⊕	⊕	⊕
	⊕	⊕	⊕	⊕	⊕	⊕

10.4.2   Other Methods of Laying

10.4.2.1 Pipe laying with crane

The common set up for laying is shown in picture below.

Photograph 14   Pipe laying with Goliath (Gantry) crane

The set up is shown below.

Photograph 15   Pipe laying in marshy land

The set up is shown below.

Photograph 16

10.5 FILLING THE PIPE TRENCH

After the pipes have been bedded, the trench is backfilled stepwise.

The soil be introduced layer-by-layer and carefully compacted up to a height of approximately 300 mm above the crown of the pipe.

If pressure testing is to be carried out in the open trench, the socket joints must also be left free of backfill until pressure testing has been completed. Further backfilling up to ground level can then be accomplished without any particular measures having to be taken. However, in the area of roads, and in urban areas careful compaction may also be necessary for this step. Trench sheeting and bracing, if any, must be removed section by section in the progress of backfilling. Since otherwise shifting earth can change the loading conditions.

When using compacting machines, particular care must be taken to avoid damaging the pipes through direct or indirect impacts.

If the pipes are protected against corrosion by a special coating, attention must be paid during lifting and compacting so that the coating will not be damaged. The bedding and backfilling material must then be selected accordingly. Any damaged places must be repaired in accordance with the instructions of the supplier.

10.6 FIELD TESTING OF CONCRETE PIPELINES

10.6.1 Introduction

Satisfactory performance of PSC pipeline do not depend only on production of required quality of pipes, but also on the correct laying and testing it, in the field. Many instances of problems in the PSC pipelines can be attributed to defective laying and not testing it. Lack of attention to design of the pipeline – both hydraulic and structural – also contributes to this.

PSC pipes are manufactured to meet the requirement of standards. Before delivery, each pipe is hydrostatically tested to a pressure specified by the user.

To ensure pipes are being correctly laid and jointed, a field hydrostatic test can be applied. Temporary bulkheads will have to be installed at the ends of the pipeline or section of pipeline to be tested.

This field test is not to be applied for the purpose of reassessing individual pipe performance. The manner in which the pipes have been treated and the conditions to which they have been subjected prior to and during laying may have affected the performance of the pipeline. It is well recognized that initial leakage may occur through sound concrete pipes, if they have been stored on-site for an extended period. Desirably such storage should be minimised by installing pipes as soon as possible, preferably before they dry out.

Pipes may show initial damp spots or weeps which will gradually diminish with time; though may not, if they are the result of damage.

10.6.2 Purpose of Test

(a) To reveal the occurrence of faults in the laying procedure, e.g., joints incorrectly installed or pipes damaged.

(b) To ensure that pipeline will resist the sustained internal service pressure to which it will be subjected.

10.6.3 Leakage Test

Perhaps the most common form of line testing is by means of the so called "Leakage Test", a term which does not clear to concrete manufacturers (It is both absorption and leakage though joints and specials). This leakage is apparent rather than real, and is a function of the absorption characteristics of concrete.

The actual test procedure requires that the pipeline, after filling, is left under operating pressure for a period of time in order to achieve conditions as stable as possible for testing. The length of this period will ideally depend upon such factors as initial permeability, movement of pipeline under pressure and quantity of air trapped.

In fact, the time allowed by the pipeline to "soak" is frequently determined by commercial considerations i.e., the desire of both the Engineer and the Contractor to complete the test in shortest possible time.

The required test pressure is obtained by pumping in water from a calibrated container, and the pressure is maintained by continual pumping. The rate of loss of water from the container is determined at regular intervals. Successive measurements will show a diminishing quantity in the case of sound and satisfactory pipeline. When the amount of water added to maintain the test pressure is equivalent to or less than the acceptable loss rate specified by the Engineer, the test is considered satisfactory.

Experience shows that short term tests at higher pressure do not necessarily reveal all defects. Of more effect is the long term test of normal operation; during which time, the pipeline settles down, bedding and back filling consolidates, and a cycle of seasonal fluctuations in climate has been experienced. Experience is that when the pipeline stands for very low pressure of say 2-3 kg/cm^2 taking it to higher pressure is a matter of time. However, if it does not stand to low pressure it is difficult to go to higher pressure.

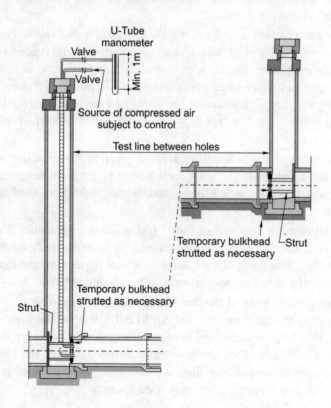

Fig. 10.11 Diagram of air test arrangement

10.6.4 Test Procedure

Length of test section depends largely on local conditions, i.e., on the size of the pipeline, on nearby structure, on soil conditions and also on the time of the year. The individual test sections should not be less than 500 m and not more than 1500 m in length. It has shown to be good practice to initially select shorter sections for testing and then gradually increase the lengths with better knowledge of the site conditions. Under such condition it is not unusual to attain lengths up to 5 km. to be tested. Lengthy sections have the advantage that the costs incurred in pressure testing are distributed over a longer section. Smaller sections are recommended when the pipeline has to be closed (recovered) as quickly as possible due to traffic requirements or bad weather.

In the course of preparing for pressure testing all horizontal and vertical bends must be adequately anchored. Straight pipe sections should also be covered to such an extent that they are adequately braced against the pipe trench wall. The pipe joints preferably remain uncovered so that inspection is possible.

Testing plates at ends may not be removed until the pipeline has been completely relieved of pressure.

Filling the pipeline must be carried out gradually and carefully so that any air present in the pipes is able to escape completely. In rubber ring jointed pipes air is trapped at socket, removing this is important. It is recommended that the pipeline be filled beginning from the low points. Between line filling and testing roughly 1 day should be interposed to permit any air present in the pipeline to escape. For this propose correct location (at the highest points) is necessary.

Considerable care and attention must be given to the provision of fixing of adequate testing plates (bulk heads) which must be properly designed to prevent blow out and leakage. Such leakage, unless monitored, can confuse the final test result.

It is desirable to conduct an initial test as soon as possible after the first 500 meters of line has been laid and jointed, as an early check on the standard of pipe laying. The feasibility of doing this will depend on the rate of laying and whether the delay to pipe laying can be accommodated.

The preconditioning of the line is important. It should be allowed to stand under some hydrostatic pressure say, up to 5.0 lg/cm^2 at the highest point on the line for as long as necessary to allow absorption of water by the concrete. Time taken will depend on the age, moisture condition of the pipes and the ambient site conditions. Some lines will need no more than 24 hours before pressurisation commences, other may need longer.

Pressurisation should be carried out slowly and can begin at 5.0 kg/cm^2 per hour. As the pipe is pressured, end plates will be thrown out due o compression of soil abutments or movements of pipes. This can be avoided by adjusting the jacks at the ends. A guide to the rate of pressure increase can be obtained at any stage by measuring the rate of less i.e., the leakage per hour, and checking how this compares with the rate given by the formula. If the rate is substantially above this there is little point in raising the pressure further at this stage. Further preconditioning should be applied if there are no obvious leaks occurring anywhere along the line.

Once the pressure has reached the specified test pressure for the line and provided no major faults have appeared the loss of water should be measured at hourly intervals, over a period of 3 hours. If these measurements show a steadily decreasing rate of leakage it indicates that the test section of line has not yet reached equilibrium. In this event it may be necessary to allow a further period of preconditioning and then repeat the measurement. The test results may be considered satisfactory when the amount of water lost in one hour does not exceed the amount defined by the formula expressed under test acceptance. I.S. 783 is silent on period of test. Present practice in India is to test for one hour but 3 hours is preferable.

10.6.5 Precautions

It is recommended that backfilling the partly completely before test is applied, to minimize the chance that pipe will 'float' in trenches which become accidentally filled with water.

It is important that a pipeline to be tested be properly restrained to prevent movement of pipes, bends, tees, junctions, adaptors and reducers. It is recommended that straight sections of pipeline be prevented from moving out of alignment by placing adequate sidefill to restrain the pipe. Excessive deflection of any pipe due to internal pressure may result in a failure at a joint.

According to German, as per article number 76 of notice number 71, this test pressure is 1.5 times the static pressure for gravity pipeline. For pumping lines, this test pressure is equal to working pressure plus the calculated value of over pressure due to water hammer plus a constant security margin of 2 kg/cm^2.

10.7 PROBLEMS ENCOUNTERED

The problems which arise during leakage test are varied. Often insufficient importance is attached to the precautions which must be taken during preparation. Bad planning and bad technique lead to bad results, and bad results lead to colossal waste of money. Many, otherwise satisfactory pipeline projects, have come to financial grief during testing.

The author's firm is involved to varying degree in the line testing of pipelines using the company products. Over 350 projects have been completed since 1965. The experience is that on some projects (contracts) where adequate thought and planning have been given before hand, to the arrangements for this line testing, completed rapidly without undue difficulties. In other cases the completion of testing has taken many months, with long delays brought about by movement of testing plated due to compression of soil, leakage through scour valves, air valves and other fittings, incorrect positioning of rubber rings, unequal compression of rubber ring due to oval spigots; and inadequate equipment for pressurising the main and measuring the water added.

Following are the more common problems encountered

(A) *Design and construction of abutments*

For testing the end thrust while testing individual sections to be adequate. It is necessary to take into account movement of end plates when the end thrust is resisted by soil. The pipes also move a little. The settlement of the soil pushes out the end plates, the plates are sometimes deflected, resulting in partially unjoint from the concrete pipe. This allows leakage past the rubber joint ring, causing eventual displacement of end plate. Many thousands of gallons of water are lost. The trench gets full of water and may endanger the surroundings. Construction of concrete anchor blocks will solve this problem, but it is expensive to construct huge concrete blocks and again dismantle them. A thought needs to be given on this subject. Following three alternatives to concrete anchor blocks, have been found to be satisfactory and economical.

(i) **Temporary anchor to resist the thrust** – can be made by laying few pipes without rubber ring and filling the sides and the top with earth, properly compacted to achieve density of about 1600 to 1800 kg/m^3. The thrust is resisted by the friction between the soil and the pipe depends upon roughness of pipe exterior. For PSC pipes with mortar coating, coefficient of friction vary from 0.25 to 0.5. Typical arrangement and worked example is given in Fig. 10.21.

(ii) **Thrust is transferred to undisturbed soil** – Trench is not excavated beyond the proposed test sections and the thrust is transmitted to the vertical face of undisturbed soil. In case the bearing capacity is not adequate, a thick steel plate is placed between plate and earth. If the thrust is still more, weights are placed on the undisturbed soil to increase its thrust resisting capacity. Typical arrangement and worked example is given un annexure 2.

(iii) **Testing against closed valves** – is very convenient, but care must be taken to ensure anchorage of valves. Valves and other fittings are notorious in giving rise to leakage through glands and packing and a careful watch must be kept on gate valve during line testing, with the amount of leakage from such fittings being measured if it is significant.

(B) *To remove entrapped air* The pipeline should be filed slowly and steps taken to ensure that air is released by correct operation of air valves. Air will be trapped inside sockets until it is slowly forced over or is carried away by water flow. Air in the pipeline is frequent cause of false readings during leakage tests and may delay tests for days.

(C) *If test is required at pressure*, higher than the available static pressure, the pumps provided must be of good standard and reliable for operating over long periods. Pressure gauges, tanks, hoses and plumbing connections must all be reliable under continuous high pressure operation. Badly maintained pumps of inadequate capacity coupled with faulty gauges are a frequent cause of delay in line testing.

(D) *Bolted and flanged joints* to steel and cast iron fittings are a frequent source of leakage, and for this reason all such joints should be left clearly visible during testing.

(E) *Rubber rings are sometimes forced out*. This is usually because of oval spigot, creating uneven annular space between socket and spigot. The ring is then compressed more in some portions and less in other than designed. Under pressure, less compressed portion of ring is forced out, causing leakage. It is therefore necessary to have sockets and spigots of correct dimensions.

(F) *Uneven seating of rubber ring*, this can be checked with feeler gauge at 3 places, when the pipe pushing equipment is still in portion and when it is removed, as shown in Fig. 10.2.

10.8 CONCLUSION

(1) Field test of a pipeline is most essential and must be conducted on all pipelines.

(2) It is prudent and economical to control the water hammer within 10-15% of working pressure when the detail hydraulic analysis is made, factors such as 1.5 or 2.0 should not be blindly applied.

(3) One should not get panic about water hammer, and make undue provisions for it and increase the cost. The maximum pressure due to water hammer can be calculated and controlled within economic range.

(4) Line test pressure should be calculated as per German Standard as indicated; to keep the cost of pipes, thrust blocks, bulk heads etc., less.

(5) During pressurising, if the joints are water tight at low pressure they rarely present problems at higher pressure; provided the dimensions of socket and spigot are within tolerance limit.

(6) Every attempt should be made to remove the entrapped air at joint.

(7) If the leakage allowance is successively diminishing, but is more than acceptable limit, it is not a matter of concern. It will diminish with time, provided rubber rings are not unevenly compressed in joints, end plates are not pushed out, abutments are not yielding and valves are not leaking at higher pressures.

(8) Abutments, which are temporary structures, to be of one of the alternatives given above, to save time and cost.

(9) Too much stress not to be given on consolidation of filling on sides and over pipe, in case of PSC pipes; as it is a rigid pipe.

(10) PSC pipe is a tailor made product. It can be economically designed for the specific pressures desired; by full utilisation of material properties. This advantage is not possible with Cast Iron and Steel Pipes. Their cost will be same for all pressures, up to 18 kg/cm^2.

10.9 TRENCHLESS TECHNOLOGY FOR INSTALLATION OF CONCRETE PIPES

Introduction

In India, Concrete Pipes for drainage, are generally installed in an open trench which is then filled with excavated material and restored to original conditions after the pipes are installed. The trenches are usually deep; and also there are many obstructions such as electricity, telephone cables etc., together with underground water. The work of installation of pipes in such conditions is difficult and time consuming. To overcome these difficulties, concrete pipes, now-a-days in overseas countries, are installed by jacking or tunneling methods of constructions where deep installations are necessary or where conventional open excavation and backfilling method may not be feasible. This practice is popularly referred to as trench-less technology. General arrangement involved is shown in Fig. 10.12. The technique broadly involves jacking horizontally the pipe itself, pressed home from a temporary working well located at end of the section to be jacked.

Concrete pipes were jacked first, in North America by the Northern Pacific rail road, between 1899 and 1900. Reinforced Concrete Pipes ás small as 150 mm dia. and as large as 3000 mm inside diameter can now be installed by this process.

10.9.1 Concrete Pipe Ideal Material for Jacking

Concrete pipe is uniquely suited for jacking application, due to its inherent compressive and shear strength. Jacking of concrete pipe avoids disruption

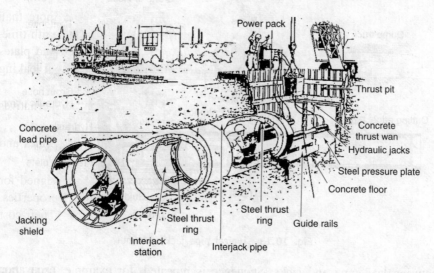

Fig. 10.12   General arrangement for pipe jacking

to traffic, either rail or road, allows installation to tight line and grades and provides a pipe with high strength characteristics to withstand any depths without buckling during installation.

10.9.2   Pipe Jacking Procedure

On the alignment of the pipeline shafts (pits) are excavated generally at manhole positions or sometimes at more distance depending upon the pipe length to be jacked. The diameter or width of the pits to be more than the length of each individual pipe. Abutments are or thrust blocks are constructed on one side of the shaft Fig. 10.13 shows the typical pipe jacking layout.

The pipe jacking process consists of transmitting a horizontal force from the vertical ground surface by means of large capacity jacks that push a concrete pipe forward at the same time as excavation is taking place, within the shield. The material at the face is excavated either by machine or manually with skilled operator who also controls the direction of the pipeline within the leader shield, under the instruction of an engineer.

The detailed operations are indicated in Enclosure-1. After the pit is excavated, abutment is constructed and jacks and frame and guide rails are fitted, first pipe is lowered and brought into position. Jacking frame and jacks are then adjusted till the pipe is pushed forwarded by jacks till it touches the face of excavation, required excavation is done in front face of pipe and the pipe is pushed further. Jacks are then retracted and spacer is added between pipe and jack. After the soil is removed through pipe, succeeding pipe is inserted

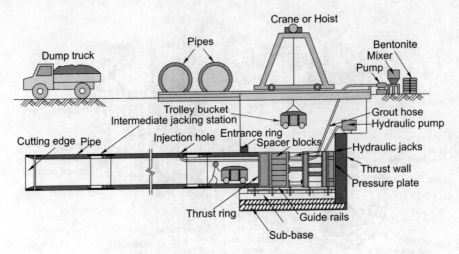

Fig. 10.13 Typical pipe jacking layout

between lead pipe and jacks. Sequence is repeated, for example, excavation, soil removal, pipe insertation and jacking until the operation is completed.

10.9.3 Suitability

Uses of pipe jacking include new sewer construction, sewer replacement and relining, gas or water mains, oil pipelines electricity and telecommunication cable installation, culvert, pedestrian subways, road carriage ways and bridge abutments. Rectangular section can also be jacked for the provision of pedestrian subways, road carriage ways culverts and bridge abutments.

New jointing methods currently being evaluated and several proprietary designs are now available. Rectangular and other special sections are normally designed to meet specific project requirements.

10.9.4 Jacking Pipe Joints

Propriety concrete jacking pipes are normally centrifugally spun precast concrete, incorporating concentric doubly reinforced cages. These are generally produced as per standard specifications but can be designed to meet more onerous requirements for superimposed loads where necessary. Standard pipes are between 1.2 m to 2.4 m long and are designed to enable the jacking forces to be transmitted along pipeline without damaging the joint. The joints normally used are shown in Fig. 10.14.

Above joints produces the required flexibility necessary for installing the pipeline and accommodating any future ground movement. The joint also provides a high degree of water tightness. Any minor leak may be sealed by caulking or other approved method.

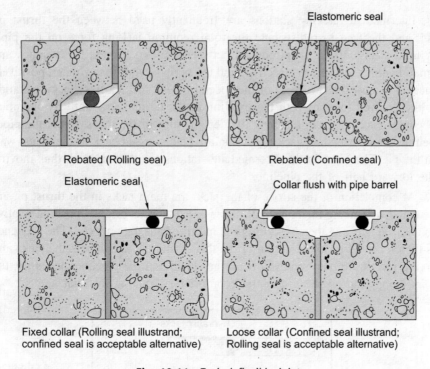

Rebated (Rolling seal)   Rebated (Confined seal)

Fixed collar (Rolling seal illustrand;   Loose collar (Confined seal illustrand;
confined seal is acceptable alternative)   Rolling seal is acceptable alternative)

Fig. 10.14   Typical flexible joints

Special lead pipes are normally produced for insertation into the shields and those contain a rebate over which the shield fits as shown in Fig. 10.15.

10.9.5   Intermediate Jacking Stations

Whenever long lengths or shot lengths are to be handled due to soil conditions, the load on the joints will very, to reduce it, interim jacking stations have to be incorporated.

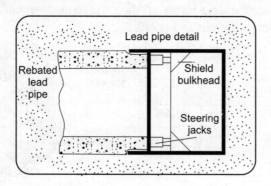

Fig. 10.15   Shield with rebated pipe

Intermediate jacking stations are frequently used between the thrust pit rig. and the face to redistribute the total required jacking force on the pipe. A mild steel cylinder is introduced between the two rebated thrust pipes and small hydraulic jacks are placed around the periphery of this cylinder, positively located against the two faces of the special pipes. The auxiliary jacking station is then moved forward with the pipeline in the normal way until its operation becomes necessary. On reaching the design value of the thrust force, the pipes behind the intermediate jacking station are held stressed back to the thrust wall in the pit. The jacks in the intermediate stations are then opened, thus moving the forward half of the pipeline.

At completion of the stroke of the jack, the main jacks in the thrust pit are actuated, advancing the latter half of the pipeline to its original position relative to the lead half and thereby closing the intermediate station jacks. The sequence is then repeated for the duration of the thrust, and on completion, the jacks and fixing pieces are then removed from the shield and the pipeline closed up, the joint being made with an acceptable proprietary jointing material.

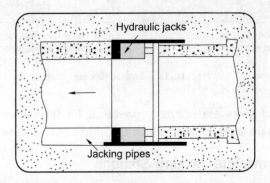

Fig. 10.16 Intermediate jacking station

Inter jack stations are not only used to increase the jacking lengths achievable but also to reduce the drag forces on the surroundings ground. Joint details of intermediate pipe are given in Fig. 10.17.

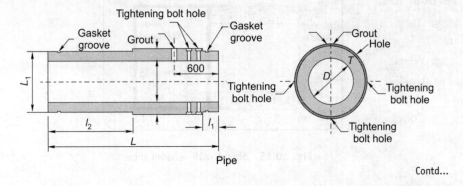

Contd...

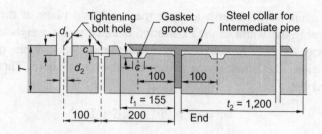

Fig. 10.17 Joint details of intermediate pipe

10.9.6 Frictional Resistance

During the pipe jacking operation, frictional forces build up around the pipeline as the line of pipes is advanced behind the shield. The frictional forces depend on the type of soil, depth of overburden, length and diameter of the pipe being jacked, the speed of excavation, whether hand or mechanical. It is difficult to accurately assess these forces, but after years of experience, pipe jacking contractors have derived empirical values. As an approximate guide, frictional forces fall between 0.5 and 2.5 tonnes per square meter of external circumferential area of pipe dependent upon site conditions and the type of excavation.

Frictional forces on the pipeline may be reduced by applying a suitable lubricant such as bentonite under pressure. If high frictional resistance is anticipated, it is recommended that intermediate jacking stations are placed at regular intervals in the pipeline.

10.9.7 Pipe Design for Jacking Pipes

Two types of loading conditions are imposed upon concrete pipe installed by jacking method. The earth load due to overburden, with some possible influences from live loads and the axial load due to the jacking pressure, applied during installation.

Vertical load on pipe

Major factors influencing the vertical earth load on pipe installed by jacking are:

- Weight of prism of earth directly above the bore.
- Upward shearing or frictional forces between the prism of earth directly above the bore and the adjacent earth.
- Cohesion of the soil.

The resultant vertical earth load on the horizontal plane at the top of the bore and within the width of the excavation is equal to the weight of the prism of earth above the bore minus the upward frictional forces, minus the cohesion of the soil along the limits of the prism of soil over the bore. This earth load is computed by:

$$Wt = C_t \, W \, Bt^2 - 2 \, C \, C_t B_t$$

Where,

Wt = Earth load lbs per linear *ft*

Ct = Load coefficient for jacket pipe

W = Unit weight of soil lbs/*cft*

Bt = Maximum width of bore excavation

C = Coefficient of cohesion of soil above the excavation in feet.

The C_t to B_t^2 term in the above equation is similar to the equation for determination of backfill load on pipe installed in the trench condition where the trench width is the same as the width of the bore. The term $2 \, C \, C_t \, B_t$ accounts for the cohesion of undisturbed soil. For cohesive soil the earth load on a jacked pipe will always be less than on a pipe laid in trench condition.

The design value for the coefficient of cohesion range, from zero to 1000, where zero is indicative of a very loose dry sand and 1000 description of a hard clay. Based on clay's large coefficient of cohesion, pipe jacked through hard clay experience little if any, earth loading even under high depths of fill.

Live loads are determined by prevailing design procedure for load distribution through earth masses for the particular transportation loading.

In general highway and aircraft load are considered in-significant at depth greater than 10 ft, however, rail loads are considered up to 30 ft of cover.

Axial load on pipe

It is due to jacking operation where pipe is pressed by jacks, the magnitude depends upon the dia of pipe, resistance offered by the soil and the length of pipeline to be jacked. It is necessary to provide for relatively uniform distribution of the load around the periphery of the pipe to prevent localized stresses concentration. This is accomplished by using a cushion material between the pipe section, such as plywood (0.5 to 0.75 inches), hardboard or similar cushion which will not compact to an incompressible material and care on the part of the contractor to ensure that the jacking force is properly distributed through the jacking frame to the pipe and parallel with the axis of the pipe.

The cross sections area of the NP3 or NP4 class as per I.S. 458 is more than adequate to resist the pressure encountered in any normal jacking operation.

Table 10.1 Indicative tonnage required for pushing concrete pipe

Outside dia of Pipe mm	Sandy Soil Tonne	Hard Soil Tonne
900	2.00	0.75
1200	2.70	1.00
1500	3.30	1.20
1800	3.90	1.50
2000	4.30	1.60
2500	5.50	2.00

Note Tonnage values are multiplied by the total length of pipe to be jacked in feet.

10.9.8   Limitations of the above process

The first problem to be handled in pipe jacking is to control the soil masses especially when clay and mud are encountered.

The second problem is to get precision without loosing speed. Lundby's solution is advanced sensors in combination with powerful steering equipment. This gives a precision down to ±0.5 cm, vertical and ±5 cm horizontal for up to about 200 m jacking distance.

The third question to be answered is how the system should be connected to other systems without disturbing the surrounding masses, i.e., without jeopardizing the groundwater level, thereby causing the soil to settle.

Fourthly, it must be possible to train key personnel without taking financial risks. This is especially important for small pipes where most, but not every, detail is automatic.

10.10   NEW METHODS OF PIPE JACKING

Today a variety pf pipe jacking methods are in use. Most of these are complicated and   expensive. Main features of important two are given below.

A. Lunday Method
B. Tele more System

(A) Lunday Method

This method overcomes all the problems mentioned in previous paragraph. The method is especially suited when the soil masses, especially when clay and mud are encountered. Firstly, by using hydraulically controlled locks in larger systems and imply by utilizing the incoming masses as counter balance in small dia pipes.

This method is especially suited for jacking below water level. A Sketch of this is given in Fig. 10.18.

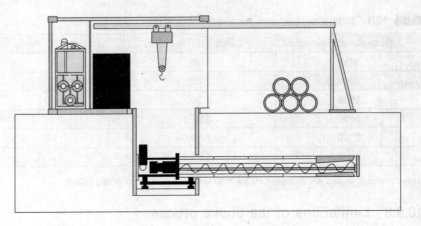

Fig. 10.18 Micro tunneling operation

(B) Tele-more System

The Tele-more is an unmanned remote controlled MEPCB type bentonite slurry shield designed for driving small bore tunnels on pipe jacking projects. Its development came about through the combination of the Iseki-developed mechanical type earth pressure counter-balanced shield and advanced remote controlled television technology. The machine has helped to eliminate many of the problems that have for years plagued small bore pipe jacking operations. As its reputation as a complete answer to the difficulties of small bore tunneling grows, the Tele-more has come to be regarded by many as the herald of a new era in safe efficient tunneling.

Detailed description of these method is beyond the scope of this paper.

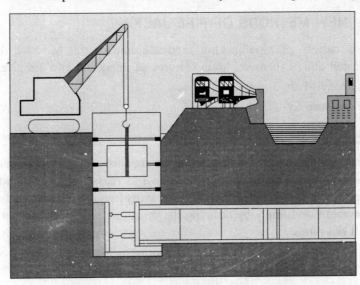

Fig. 10.19 Tele-more system – Jacking pipe below water

10.11 JACKING PIPES FOR BAGHDAD STORM WATER DRAINAGE PROJECT (IRAQ)

The Indian Hume Pipe Co. Ltd., manufactured the following pipes for Baghdad storm water drainage project.

Total length of pipeline	56,134 Meters
Numbers of pipes	23414 Nos.
Total wt. of pipe	1,12,113 M.T.
Total period for production	18 Months
Length of each pipe	2.4 Inches
Dia. in mm varying from	450 to 2500
Out of the pipes for jacking	2436 Nos.
Year in which work was done	1980-83

Details of the pipes is given below.

Table 10.2 Details of Jacking pipes including reinforcement

Pipe I.D.	Thickness	Reinforcement				Length of pipe	Three edge load
		Circumferential		Longitudinal			
		Nos	Dia	Nos	Dia.		
mm	mm	mm	mm	mm	mm	mm	Kg/m
1000	115	62	4.5	12+12	4.5	2430	5200
1500	150	53	6.0	12+12	6.0	2430	7450
		70	6.0				
1600	165	43	6.0	12+12	6.0	2430	8200
		38	6.0				
2000	190	45	7.0	12+12	7.0	2430	9700
		60	7.0				
2250	215	46	8.0	12+12	7.0	2430	11200
		62	8.0				
2500	240	34	8.0	16+16	7.0	2430	12700
		46	8.0				

10.12 DETAILS OF SPECIAL FITTINGS

Special fittings such as bends, tees and others supplied with Prestressed Concrete Pipes are made of steel shells with a concrete lining and outer coating. The ends are fitted with rubber ring joints for connecting them to the concrete pipe.

Dimensions of Bends							
Nominal Internal Dia. MM	61° to 90° R	Bends L3	46° to 60° R	Bends L2	31° to 46° R	Bends L2	30° Bends L1
600	660	990	660	690	990	690	530
700	760	1140	760	840	1140	840	610
900	910	1300	910	910	1370	910	610
1100	1070	1600	1070	1140	1600	1070	760
1200	1220	1680	1220	1220	1830	1140	840
1400	1370	1830	1370	1220	2060	1220	840
1500	1580	2030	1580	1370	2290	1370	910
1600	1680	2130	1680	1370	2520	1370	910
1800	1830	2210	1830	1450	2740	1450	910

Dimensions of Air Valve Tee and Scour Valve Tee		
Nominal Internal Diameter in mm,	Dimension "C"	Clearance "D"
301 to 700	350	100
701 to 900	450	150
901 to 1800	600	150

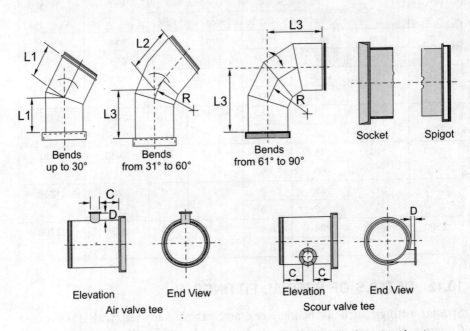

Fig. 10.20 Dimension of specials from from Air valve tee and scour valve tee

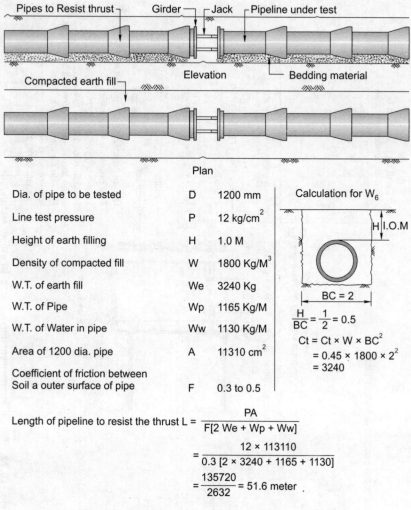

Pipes to Resist thrust — Girder — Jack — Pipeline under test

Elevation — Bedding material

Compacted earth fill

Plan

Dia. of pipe to be tested	D	1200 mm	Calculation for W_6
Line test pressure	P	12 kg/cm^2	
Height of earth filling	H	1.0 M	
Density of compacted fill	W	1800 Kg/M^3	
W.T. of earth fill	We	3240 Kg	
W.T. of Pipe	Wp	1165 Kg/M	BC = 2
W.T. of Water in pipe	Ww	1130 Kg/M	$\dfrac{H}{BC} = \dfrac{1}{2} = 0.5$
Area of 1200 dia. pipe	A	11310 cm^2	$Ct = Ct \times W \times BC^2$
Coefficient of friction between Soil a outer surface of pipe	F	0.3 to 0.5	$= 0.45 \times 1800 \times 2^2$ $= 3240$

Length of pipeline to resist the thrust L $= \dfrac{PA}{F[2\,We + Wp + Ww]}$

$$= \frac{12 \times 113110}{0.3\,[2 \times 3240 + 1165 + 1130]}$$

$$= \frac{135720}{2632} = 51.6 \text{ meter}$$

Number of 5 m long pipes required to resist thrust $= \dfrac{51.6}{5} = 10.3$ pipes

Say II pipes

Length of pipeline (Laid without rubber rings) to resist thrust

Fig. 10.21 Typical set up for resisting thrust during line testing

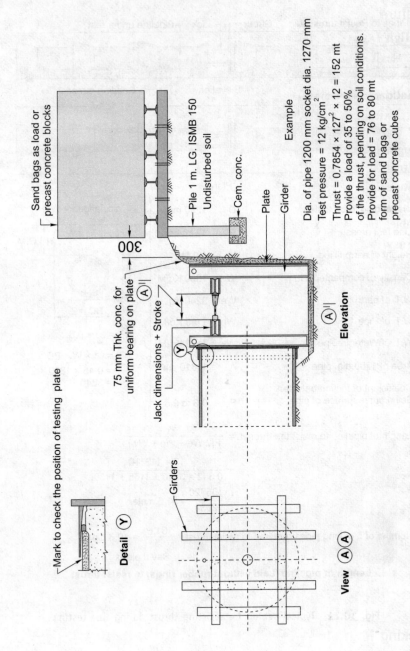

Sand bags as load or precast concrete blocks

Pile 1 m. LG. ISMB 150

Undisturbed soil

Cem. conc.

300

75 mm Thk. conc. for uniform bearing on plate

A||

Jack dimensions + Stroke

Y

A||
Elevation

Plate

Girder

Example

Dia. of pipe 1200 mm socket dia. 1270 mm

Test pressure = 12 kg/cm²

Thrust = 0.7854 × 127² × 12 = 152 mt
Provide a load of 35 to 50% of the thrust, pending on soil conditions.
Provide for load = 76 to 80 mt form of sand bags or precast concrete cubes

Mark to check the position of testing plate

Detail Y

Girders

View A A

Fig. 10.22 Typical arrangement for resisting thrust at factory site

10.13   FLEXIBLE PIPE

Feature

1. High Degree of Flexibility

The flexible part of pipe permits deformation and easily absorbs bending stresses and shear forces, and is therefore superior in adaptability.

2. Solid Construction

The flexible part is made integrally with the pipe wall. It has been centrifugally cast and is steel reinforced for construction. For this reason, it has a unit strength greater than that of the pipe body itself, and is better able to resist external forces.

3. Superior durability

High quality synthetic rubber is used in the flexible part which is extremely durable, and anti-chemical and oil-proof.

4. Easy to Use

Since there is no change in the shape of the pipe, and dimensions and elbow construction of Hume concrete pipe are standard treatment and installation can be performed in the normal way, and requires no special machines, tools, etc.

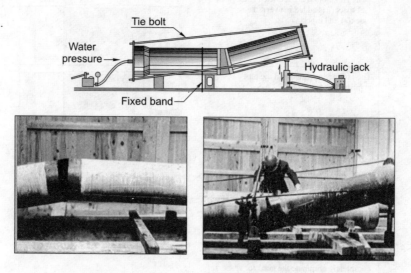

Fig. 10.23   Method of test of flexible Hume pipe

Jacking Reinforced Concrete Pipes

Normal pipeline construction methods – excavating, pipe laying, Backfilling –are not always practicable. This becomes evident when considering pipelines beneath existing highways and railway lines.

A practical alternative is accepted method of jacking, and Humes manufacture suitable pipes for this purpose.

Step in Jacking Concrete Pipes

Pits are excavated on each side. The jacks will bear against the back of the left pit so a steel or wood abutment is added for reinforcement. A simple track is added to guide the concrete pipe sections. The jack(s) are positioned In place on supports.	Jack — Abutment — Track — Jack support
A section of concrete pipe is lowered into the pit. This pipe may be fitted with a cutting edge or a shield (full or partial) if warranted by ground conditions.	Pipe
The jack(s) are operated pushing the pipe section forward into a heading usually excavated from within the pipe.	
The jack ram(s) are retracted and a "spacer" is added between the jack(s) and pipe.	Spacer
The jack are operated and the pipe is pushed forward again.	
It may be necessary to repeat the above steps four and five several times until the pipe is pushed forward enough to allow room for the next section of pipe. It is extremely important, therefore, that the stroke of the jacks be as long as possible to reduce the number of spacers required and thereby reduce the amount of time and cost. The ideal situation would be to have the jack stroke longer than the pipe to completely eliminate the need for spacers.	Pipe Pipe
The next section of pipe is lowered into the pit and the above steps repeated until the operation is complete.	Pipes

Joint Details	
Flush joint pipe are readily installed by jacking, using a suitable end packing such as rope which is fitted around the male end of the pipe. However, if the jacking force is excessive, the eccentric transfer of such force from pipe to pipe results in a bending action on the pipe wall which can lead to fracture. Humes staff will be pleased to advise in cases of doubt.	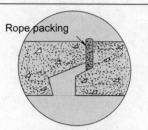
Butt joint pipes provide a large contact area for high jacking forces and also eliminate the eccentricity. These pipes require a steel locating band as shown in the detail.	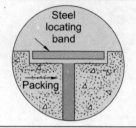

- Concrete Pipe for Jacking - Pipe Jacking Association, London - August 1993.
- Jacking Concrete Pipe - The Concrete Pipe Association, London - February 1991.
- Pipe Jacking in Soft Soil - LUNDBY - Marland International, Vendevagon 63, 18264 Diursholm Sweden.
- Decond Emergency Wylds Road, Bridge Water Somerset TA 64 BH U.K.
- Microtumullery - Reprint from underground - September 1989.
- TELE - MOLE - Ireki Poly - Tech, Ltd., 23.3 Ichibanco, Chiyad….. Tokyo 102 Japan.
- Reinforced Concrete Pipes for Jacking Installation - Japanese Sewage.
- Works Association JSWAS - A - 2 - 1975.
- ACPA

CHAPTER 11

Distributed Reinforced Concrete Pipes

INTRODUCTION

A novel large-diameter reinforced micro Crete pipe, called distributed reinforcement pipe (DRP) is being produced by the Vianini Company in Italy. It has been used successfully in pipelines from 600 to 1400 mm (24 to 55 in.) in diameter at working pressures from 600 to 1200 kpa (90 to 170 psi). The pipe is manufactured by projecting a rich micro Crete over a rotating mandrel while simultaneously placing both circumferential and longitudinal thin wire reinforcement.

The thin steel wires used to reinforce DRP usually are 0.8 or 1.0 mm (1/36 or 1/25 in.) in diameter, and are placed along the pipes two main stress directions, circumferential and longitudinal (see Figs. 11.1 and 11.2). This produces a highly efficient orthotropic material, as with long fiber composite materials, compared to fiber reinforced concrete with randomly distributed short fibers. In the latter case, the random distribution gives an efficiency of about one third, without considering the loss of resistance due to the interruptions of the short fibers.

Although similar to Ferro cement, DRP is actually a special type of reinforced concrete with high values of resistance, especially in traction obtained through the high quality of its components as well as the fine disperse of reinforcement. In Ferro cement, the resistance of the material is supplied almost exclusively by the steel cage, which forms an almost self-sufficient structure, and the concrete mainly holds together and protects the steel mesh.

Because the distribution of reinforcement in DRP follows the stress demands of pipe loading as it is applied in traditional reinforced concrete and good quality hand-laid composite materials, circumferential reinforcement is by far the more important, ranging from 1% to 5 or 6% in cross section. Longitudinal wires usually are necessary only up to 0.5%, depending on fill conditions and overall pipe diameter.

11.1 MATERIALS

The concrete for DRP is made from crushed limestone and river sand, both washed, depulverised, and sieved, and with a maximum particle size of about 3.5 mm (0.14 in.). Cement content varies between 500 and 700 kg/m^3 (840 and 1180 lb/yd^3) by weight. The water-cement ratio ranges from 0.28 to 0.32.

Mechanical characteristics of the concrete have been determined using specimens cut from un-reinforced rings of concrete projected on a mandrel and steam cured. Direct tension specimens were cylindrically shaped, with two terminal expansions to accommodate the testing grips without altering the proper stress field. These specimens were machined from hardened concrete on a lathe or rotating grindstone. Prismatic specimens were also cut for comparing results with those of other tests, and for bending and elastic modulus tests. A few compressive strength tests were also made, although this value is not important for this product.

The relevant values are as follows:

- Direct tensile strength (37 specimens) $f_{dt} = 6.3$ Mpa (920 psi); standard deviation $s = 1.46$ Mpa (230 psi)
- Bending tensile strength (9 specimens)—$f_{bt} = 10.69$ Mpa (1550 psi); $s = 0.82$ Mpa (119 psi); $f_{bt}/f_{dt} = 1.59$
- Compression strength (4 specimens)—$f_c = 69.0$ Mpa (10,000 psi); $f_c/f_{dt} = 10.9$
- Tensile modulus of elasticity — $E_c = 37,300$ Mpa (5,400,000 psi); $s = 1.23$ Mpa (178 psi); $E_S/E_c = 5.63$

Drawing hardens the wire to the point that it behaves linearly up to maximum strength, at values ranging from 800 to 900 Mpa (116,000 to 130,000 psi)

Reinforced tensile specimens were tested to determined composite behavior (steel-concrete bond), but, because of the reduced length of the specimens, failure was caused by the slipping of the wires instead of the breaking of steel. Nevertheless, the steel-concrete bond had an average value of 3.0 Mpa (430 psi).

11.2 MANUFACTURING

Concrete with a water content of about 25% is prepared in a mixer, then conveyed by a belt to the projecting device, which consists of two counter-rotating wheels that throw the mix against the mandrel at a speed of about 28 m/sec. (over 60 mph) (see Figs. 11.3 and 11.4). For ease of operation, the rest of the water is sprayed onto the concrete 90° from the projecting point while the mandrel turns.

The mandrel, a steel cylinder, is coated with paraffin (EPA approved) to ease freeing of the hardened pipe. It turns at 50 to 30 revolutions per minute, depending on pipe diameter. Two pipes are laid at a time, with the belts at the middle of the mandrel. The circumferentialreinforcement is wound onto the mandrel in 180 mm wide (7 in.) bands containing 16 to 58 wires, depending on the design.

The reinforcement bands are composed during pipe construction, with the single wires pulled from coils and assembled by passing through templates. The longitudinal reinforcement is formed in zig-zag "gauze" 250 mm (10 in.) wide and is held together by two nylon threads (see Fig. 11.5).

The mandrel moves from side to side along its longitudinal axis while the projecting and winding process takes place. The pair of pipes, still on the mandrel, are then transferred to a steam tunnel for approximately 3 hours curing at 50 C (132F). After this, the two pipes are cut apart and taken off the mandrel, then further cured in a second steam tunnel for about 6 hours.

11.3 LIMIT STATES

In designing concrete structures, say frames or plate, both yield and ultimate limit states are critical conditions that must be considered. The first critical situation occurs when the reinforcement yields, this is being the threshold beyond which large, irreversible deformations take place, putting the structure out of commission; the final stage is the collapse of the structure. The corresponding loading conditional can be roughly defined as the one at which the structure offers its maximum reaction, which is equitant to when any p/d curve enters a softening branch. Of course, these definitions are merely theoretical and must be adjusted, taking into account the peculiarities of the different classes of structures and loading history.

We think those concepts can be adapted for reinforced concrete pipes where an equivalent of the yield point is shown by the first appearance of moisture on the pipe wall. Failure occurs when the leaks are such that the pipeline no longer functions. This later situation can be compared to a softening branch in mechanically loaded structures, because it can be said that the load has reached its maximum value, at least up to the largest flow that can be supplied by the pumps.

Considering the specific problem of pipes this way, the usual dry testing and even the design formulas (e.g. ASTM C 497, gage leaf for measuring cracks, or DIN 4035 standards) could appear just as effective and good shortcuts for evaluating the actual behavior and safety of the product at hand.

Considering the novelty of the DRP and its particular behavior, a rather complete set of wet tests was performed during development of the pipes about 15 years ago, and then repeated systematically later.

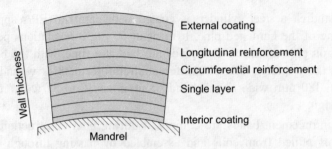

Fig. 11.1 Longitudinal and circumferential reinforcement

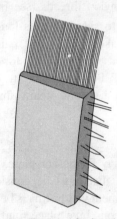

Fig. 11.2 Distribution of wires in a pipe wall segment

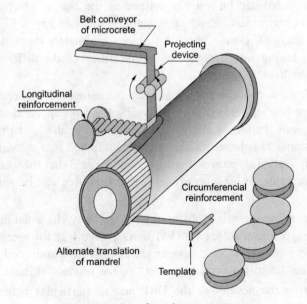

Fig. 11.3 Manufacturing apparatus

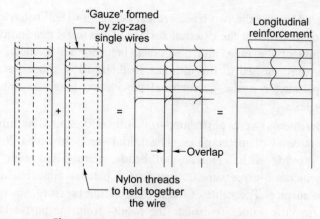

Fig. 11.4 Longitudinal reinforcement gauze

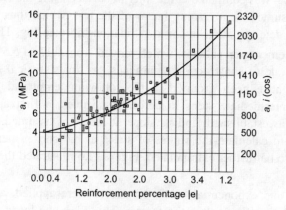

Fig. 11.5 Interpolating procedure for statistical analysis

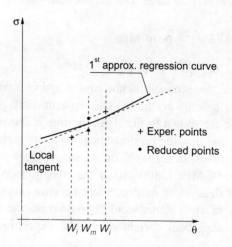

Fig. 11.6 σ-ρ regression curve

The tests addressed the two basic critical periods in DRP behavior: the first appearance of moisture at the external surface of the walls that indicated initial micro cracking of the pipe, and post-crack behavior up to massive leakage. A few tests were pushed to the ultimate load. In all of those tests, the only significant parameter is the geometrical percentage p of the reinforcement. First-loading tests

The experiments were performed on 500, 600, and 800 mm (20, 24, and 31 in.) internal diameter pipes. The initial series involved 76 tests. The specimens, provide with adequate bulkheads, were filled with water and subjected to increasing pressure until moisture patches appeared all over the external pipe surface. The value of the corresponding pressure was taken as the first micro cracking value. To isolate the results from the particular diameter and thickness of the single test, the value of the unit stress σ_1 – initial micro cracking stress's – was computed at the interior surface using the Lame formula, and there by assumed as governing factor of the test, which thus assumed the meaning of the determination of the σ/ρ relationship (see Fig. 11.6)

The reinforcement percentages, which ranged from 0.17 to 5.5%, happened to be scattered due to construction techniques. So, for statistical extrapolation purposes, they have been grouped in the following empirical way: having chosen a few intermediate steps in percentage values, a first approximation best-fit curve was established, and its slope at the steps assumed. In this way, it was possible to calculate the values of stress at the chosen percentages for the single tests to be further extrapolated. Figure 11.6 illustrated the geometrical procedure.

Both linear and exponential best – fit curves were applied, and the latter was found more significant than the former. This is shown by the lesser value of the mean square in the second case. The exponential best fit curve follows the law

$$\sigma_1 = 3.73 \; e^{0.267} \; \rho, \text{ in Mpa}$$

$$\sigma_1 = 530.5 \; e^{0.267} \; \rho, \text{ in psi}$$

For practical purposes, the exponential law can be approximated by simpler parabola with negligible loss of accuracy. The experimental points and the exponential best-fit curve are shown in Fig. 11.6. the figure shows that the pipe behaves as a composite, and that the reinforcement is of the highest efficiency, more than proportional to its percentage, up to surprisingly high values of tensile strength – nearly 16 MPa (2300 psi) at the maximum ρ of 5.4% It is difficult to obtain higher degrees of reinforcement because projecting concrete through the thick layers of steel wires would compromise the quality of the hardened concrete. The statistical correlation of the experimental values is rather good.

Four additional series have been tested, totally 49 pipes of the same diameters, to investigate the effect of some improvements in fabrication technology and reinforcement pattern that had been introduced. The results show a very good agreement between the two groups.

11.4 BEHAVIOR AFTER INITIAL MICRO CRACKING AND REPEATED LOADING

The property of cracked concrete pipes to self-seal small fissures when reloaded in the presences of water is well known. Presumably, the phenomenon is due to the hydration of free cement when water seeps through the crack. This behavior is magnified in DRP because of the very high cement content of the mix, which certainly has not completed its chemical reaction. If the pressure is maintained for a sufficient time and then released, the increase in volume that accompanies this chemical reaction causes a permanent state of stress, with the steel in tension and concrete in compression.

The mechanism acts in two ways: the tensile strength of concrete, initially lost across the cracks, is partially restored due to the glue effect of the scaling, and the prestressing raises the pressure level at which the concrete is put in traction. The actual effect has been investigated by systematically retesting many pipes in the latest series. During retesting, a higher value of incipient leakage pressure was obtained; the ratio of he corresponding tension σ_2 to the first-loading value σ_1 ranges between 1.4 and 2.6 (see Fig. 11.8); the scatter is due mainly to some differences in reloading history during the series of tests.

Figure 11.9 shows the reduction in leakage of a pipe initially micro cracked end then maintained at the design pressure, plotted against time. It appears that the sealing effect is obtained in a short time. Note, however, that the high value reached in the steel stress is not far from the yield point: it follows that high grade steel must be used to maximize the benefits of this type of pipe. Due to these practical difficulties, only a few conventional ultimate load tests were performed in which the yielding of reinforcement was reached.

11.5 TEST RESULTS AND DESIGN CRITERIA

In conclusion, the tests indicate three main ranges of reinforcement percentage with significant differences of behavior. At very low values of ρ, that is less than 0.5%, the distributed reinforcement effect is almost negligible: the pipe reacts almost as a plain concrete pipe, that is, rather fragile and without any appreciable self sealing.

At a high values (e.g. over 4.5%), the behavior is more ductile, the self-sealing is sure and evident, but the maximum tension in the reinforcement can be close to the yield point, so it is important to ascertain the relevant degree of safety. In the intermediate range, in which most of the applications occur, the self-sealing is reliable, and there is a large safety margin against yield.

Thus, design procedure is reduced to computing the maximum tensile stress under circumferential moments due to external load and internal pressure, and comparing it to the admissible value given by the σ/ρ function. Note that the term due to circumferential bending is reduce by the factor 0.6. This fact depends on two similar but distinct circumstances: the first one, already suggested by Guerrin[2.3] considers the higher strength obtained from bending than from pure tension test on plain concrete, while the relationship σ/ρ is obtained under internal pressure, that is pure tension. The second one depends on the fact that, in combined tension and bending, especially when the eccentricity is high, there is an appreciable gradient in the tension distribution along the wall thickness, and some layers can be in compression. Such situations can have a significant effect on micro cracking and subsequent seepage of water.

A more sophisticated approach would use an interaction moment/pressure curve – not too different in concept from Spangler's parabolas – into which a proper correction must be introduced to take the gradient effect into account. Further testing is needed, as described in the following, to obtain reliable experimental values allowing the use of such curves.

11.6 EUROPEAN CODE – ACCEPTANCE TESTS

There is a proposal to include DRP in the CEN (European) code for pipes. This is currently being prepared, in the sub clause regarding reinforced concrete pipes in general. In this code among other subjects, the specifications concerning wire position are as follows: "the specific volume of circumferential wire shall be no less than 1%. In each (of a minimum of six) layer, the design spacing between adjacent wires shall be not less than three times the wire diameter and not more than twice the largest size of aggregate used; deviation from design spacing shall be accepted if, in a longitudinal section of the central part of the barrel, the following conditions are met: five adjacent squares having sides equal to the wall thickness, contain a total of at least 90% of design number of wire".

An analogous clause concerns longitudinal reinforcement, with a minimum of 0.2 percent in specific volume. The acceptance criteria for DRP concern the usual factory and field pressure tests as well as crushing and bending.

Due to the peculiar behavior of DRP, crushing and bending tests are performed in combination with internal pressure (wet tests). The crushing pressure apparatus is described in Fig. 11.10; the loading system is composed of a beam having appropriate stiffness, loaded at four points to obtain as uniform a loading as possible all along the pipe barrel. This aspect is of great importance:

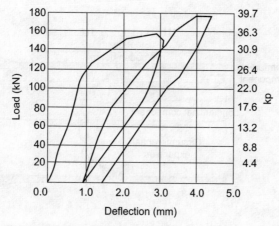

Fig. 11.7 Linterpolating procedure for statistical analysis

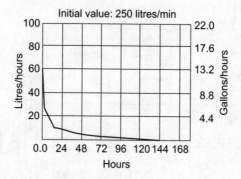

Fig. 11.8 Times plotting of leaks in a cracked pipe

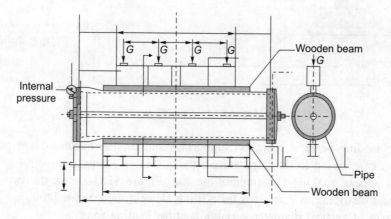

Fig. 11.9 Crushing test

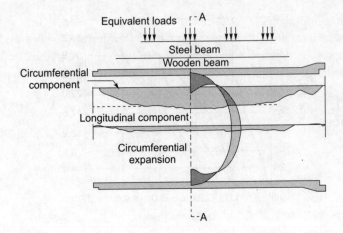

Fig. 11.10 CFiE analysis of crushing test

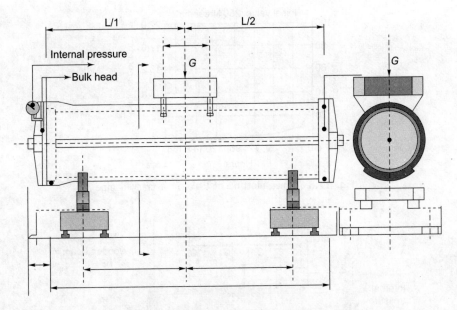

Fig. 11.11 Bending test

if the loading beam is too short, important longitudinal bending takes place; in this case owing to the large anisotropy in reinforcement, transversal cracks could appear, making meaningless the test. Figure 11.11 shows the two-way moment distributions computed for a finite element model of a 1400 mm pipe in a plot of vertical diameter variation against loading force.

The change in slope that marks the start of micro cracking and the appearance of significant moisture patches at the surface is evident.

Longitudinal strength is checked through a three-point bending system together with internal pressure (see Fig. 11.12). The testing devices are currently being employed to refine knowledge of two-way behavior, thus defining the physical parameters that are necessary to establish an improved interaction curve for design.

References

"Ferrocement – Materials and Applications", SP–61, American Concrete Institute, Detroit, 1979, 204pp.

Guerrin, A., "Traite de Beton Arme", Tome I, Paris, Dunod, 1959.

Guerrin, A., and Daniel, G., "Le Calcul des Tuyeaux en Beton Arme et non Arme. Paris, Eyroller", 1952.

Concrete International May 1993.

CHAPTER 12

Performance of Reinforced Concrete Pipes

SCOPE

The broad subject of performance of reinforced concrete pipe can, at best, only be superficially reviewed in a short paper such as this, so after a brief look at the basic physical and chemical properties of reinforced concrete, attention will be directed primarily to

- The performance of R.C. pipe in typical underground installation, where Australian case histories going back over 50 years are available.
- The fairly wide range of conditions where concrete pipes have special advantages over other types of pipe.
- The type of installation conditions where use of reinforced concrete requires special consideration – for example, where the soil environment may be aggressive to concrete.

12.1 BASIC CONCRETE PROPERTIES

12.1.1 Chemical Composition

Concrete normally consists of inert aggregate bound together by hydrated cement and most of the fundamental concrete properties derive from the chemical properties, although mix proportions, the degree of compaction or consolidation achieved when the concrete is placed and subsequent curing can have important modifying effects.

In simplified terms normal Portable cement can be considered essentially as a mixture of tricalcium silicate C_2S, dicalcium silicate C_2S, trecalcium aluminates C_3A, tetra calcium aluminoferrite C_4AF with some gypsum $CaSO_4$ added to control setting times. When mixed with water a complex hydration reaction occurs forming complex calcium silicate hydrates, calcium aluminates hydrates and calcium hydroxide, or lime. These complex hydrates form the binder commonly known as the "cement gel" – which is a rigid gel consisting of an extremely fine interlocking crystalline structure. The hydrated calcium silicates are the most important of the cementing compounds.

Australian standard AS 1315-1973 specifies four types of cement and all are used in concrete pipe manufacture. Whilst Type A (normal) and Type B (high early strength) are most common, Type D (sulphate resisting cement) is often used to provide additional durability In "sulphate" conditions. Some sulphates react with calcium aluminate hydrate to form calcium sulpho – aluminate within the concrete. This expansive reaction can cause disintegration but, in general, it will not occur with Type D cement where the C_3A is limited to a maximum of 5%. Type C (low heat) is not commonly used in pipes.

12.1.2 Physical Properties

The general principles of concrete technology are applicable to pipe but any general assessments of pipe concrete should be made in the context of the usage of extremely low water cement ratios and very high comp active forces. In literature relating to reinforced concrete one seldom sees reference to water cement ratios below 0.40 by weight, but this is near the upper limit for concrete pipe and water cement ratios significantly below this are the norm. These low w/c ratios are only practicable because of the high comp active forces possible in the centrifugation, vibration and pressure methods used by pipe manufactures.

These factors result in high strength concrete – 70MPa compressive strength is common – and this is in the range now being researched in structure concretes as "Ultra high strength concrete".

Strength gain continues significantly for long periods if moist conditions pertain. Pipe dimensions are stable.

12.1.3 Permeability

Low permeability is one of the most important properties of pipe concrete, not only because it is an essential feature of an hydraulic pipe, but most importantly, it is an indicator of concrete durability.

Permeability is primarily controlled by the final water cement ratio of the mix and the compaction and curing of the concrete. The hydration process stoichemetrically requires a water cement ratio around 0.24 to 0.28 by mass for full cement hydration. The w/c ratio required for work ability exceeds this figure and this can result in voids in the hardened concrete. Failure to properly compact the concrete also produces voids. The effect of water cement ratio on permeability is shown in Annex I – w/c ratio below 0.40 can give virtually "impermeable concrete.

The hydration process is a relatively slow one so that many of the capillary pores existing with in the matrix at an early age will be eliminated by being filled with hydration products as the reaction proceeds in the presence of water.

Many cases of chemical deterioration of underground concrete can be attributed to concrete permeability – the destruction due to wetting and drying, reinforcement corrosion, more rapid chemical attack, etc. results from moment of liquid into or from the concrete. Properly made concrete pipe has low permeability because of the dense concrete with low w/c ratio, and result is excellent durability.

12.1.4 Reinforcement

Pipe design and structure function served by reinforcement is the subject of another paper, but the conditions established which prevent steel corrosion are worth mentioning.

In the hydration process the limit liberated creates a highly alkaline environment, the pH produces a state of passivation of the steel – a series of electrochemical reactions increase the electrochemical potential of the steel surface – the iron oxide film formed in effect "enables" the steel and makes it behave as a corrosion resistant metal.

Cover to reinforcement is important in maintaining the high pH to protect the steel – a permeable concrete can allow leaching of lime or ingress of aggressive. Reference to Fig. 11.1 shows that a w/c ratio change from pipe concrete (0.400 to builders structure concrete (0.65) can increase the permeability several hundred times. This explains why the thin wall reinforced pipe sections with their relatively low cover have not experienced steel corrosion and consequent spilling as have pores concretes with much greater cover.

12.1.5 Cracks

The durability of concrete can be affected by cracking, which may arise in a number of ways – shrinkage in the cement paste, physical damage in handling, thermal stresses or structure overload. Shrinkage cracks are probably the most dangerous as they may contribute to overall porosity – adequate water curing is most important factor in minimizing this effect. Pipes, like all reinforced concrete, are designed to crack under load as the full strength of the steel is not utilized until this occurs but steel stresses are limited to keep crack width low.

The crack width is the important factor in determining whether or not the cracked concrete has been made more permeable. It has been established water has no mobility in cracks of less than 0.1 mm ("0.004") width, so that these cracks do not allow the ingress of aggressive solution. Australian Pipe standard limit crack width at the pipe surface, it must be remembered that crack width at the reinforcement steel are very much less.

12.1.6 Autogenous Healing

Concrete which is cracked or porous and showing permeability will often heal itself in the presence of moisture by the process known as autogenous healing. The water passage is sealedby one or more of the following.

- Swelling of the cement gels due to absorption of water (the inverse of the shrinkage effect)
- Recommencement of the hydration of hitherto unhydrated cement particles.
- A reaction between the free lime in the concrete and carbon dioxide either from the atmosphere or dissolved in the water. This reaction produces insoluble calcium carbonate which can plug up the voids.

Autogenous healing of reinforced concrete allows pipes which are overload and cracked to again become watertight when the overload is removed, or in the case of internal pressure to heal whilst still under working pressure.

12.1.7 Effect of Modern Technology

During the last 50 years a variety of pipe manufacturing processes have been developed, with a resultant increase in pipe quality, longer lengths and ever increasing pipe diameter. (The spun and rolled processes are Australian inventions, and have been widely licensed overseas – a recognition of the Australian quality).

12.2 PERFORMANCE OF REINFORCEMENT CONCRETE PIPE

In many ways the ideal exposure situation to ensure maximum concrete durability is underground. Soil moisture ensures continued curing and strength gain, temperatures are relatively constant minimizing thermal stresses and in most cases, drying shrinkage cannot take place. Most soil is not aggressive to concrete.

The classic examples of concrete pipe durability go back to Roman days where both sewer and water supply used concrete pipe lines. A short length of line constructed by the Romans is supplying water to the city of Trier in Germany; a Rome sewer is still in operation.

Australian service history cannot go back 2000 year, but there is considerable evidence of satisfactory performance of R.C. pipes for over 50 years. Some of the early concrete pressure pipe contracts in Victoria, prior to 1920 were.

Traralgon 4160 m (13,662') of nominal 250 mm (10") diameter with rest pressure ranging from 300 kPa to 70 kPa ('100' to 250' head).

Kerang 1875 m (6148') of nominal 225 m (9") diameter with 450 kPa ('150' head) test pressure.

Mitcham 3420 m (11,220') of 100mm (24") diameter with rest pressures ranging from 240 to 480 kPa (80' to 160' head) and 220m (720') of 90 mm (30') diameter to 250 kPa pressure (85' head). and similar pipes, totally nearly 30500 m (100,000') for Kerang, Wangaratta, Mitcham, Flinders Naval Base, Swan Hill and Corryong. Pressure pipes were also known to be installed prior to 1920, in Queensland, South Australia, Western Australia and Tasmania.

As these lines are still in service it is difficult to recover and test theses old pipes. However, a pipe was recently recovered from a SR and WS commission Dromana-Portsea pressure line – 375 mm (15") diam. – made prior to 1940 and found by test to still meet the specified requirements.

An example of the gain of pipe strength in service arose in 1964 from a decision by the violet Town waterworks thrust to increase the operating pressure of the local water supply to 600 kPa (200' head) – pipes 150 mm (6") diameter were supplied in 1926 with rest pressure 540 kPa (180' head). Several of these original pipes were taken from the line and pressure tested, showing failures at pressures in excess of 1350 kPa (450' head) thus giving confidence for a problem free operating pressure increase, a surmise subsequently proven correct.

It is of interest to note that the some authorities require pressure pipe to be permanently identified prior to laying so that, when time renders the line obsolete and the pipe is relaid its design performance can be readily ascertained.

In the sewer pipe field similar satisfactory performance over a 50 year period can be validated. A 1050 mm (42") diameter factory made pipe was laid through rock fill to discharge sewage into the sea, near Hobart, in 1921. in addition to the sewage, the pipe was exposed to sea water both internally and externally. Although theses pipe were made by methods now considered antiquated (for example, the reinforcement was hand made tied, not welded, and the joint was made by a separate concrete collar) the concrete showed no sign of determination and the reinforcement was in as new condition, when the pipe was examined in recent years.

For example, load there is a high factor of safety between the 0.15 mm (0.006") crack design load and the actual load applied in the field service. No case of structural failure of quality pipe laid as designed has been reported throughout Australia.

12.3 SPECIAL ADVANTAGES OF REINFORCEMENT CONCRETE PIPES

These are a number of laying situations where reinforced concrete pipe has technical advantages over other conventional pipe materials, irrespective of cost factors. These situations include.

12.3.1 Clay Soils

In many clay soils, typified by the Horsham and Geelong areas in Victoria, changing soil moisture content causes significant movement which, unlike settlement, persists through the life of the pipeline. As a result a history of failure of 3.96 m (13') long fibro pipes has been built up, due to their having insufficient beams strength. With reinforced concrete pipes in 1.68 m (6') or 2.44 m (8') length, and flexible joints, the beam strength is adequate to resist these soil loads.

12.3.2 Effect of Overload

Reinforced concrete pipe will remain serviceable even if subjected to occasional or accidental overload – for pressure pipe the autogenous healing affect has already been mentioned. For reinforced concrete pipe under external load an overload of up to nearly 50% will not cause disastrous failure and, if the overload is removed recovery will occur, usually leaving the line serviceable. Non reinforced pipes are destroyed in these circumstances.

Further, severe handling damage which causes destruction of non reinforced pipe results only in chipping of reinforced concrete, easily repairable with modern techniques like the epoxy resins.

12.3.3 Seawater Exposure

Reinforced concrete pipes have a proven history of satisfactory service in marine exposure conditions. Whilst sea water is basically a solution of sodium potassium and magnesium sulplates and chlorides and hence appears to create an aggressive sulphate condition, the presence of the chlorides modifies the sulphoaluminate reaction, and chemical corrosion of the concrete is of no consequence. Corrosion of the reinforcement is prevented by dense concrete although cover is sometime increased as a safeguard against chloride attack on the reinforcement. Sulphate resistant cement could be a extra precaution.

12.3.4 Dimensional Stability

Concrete kept moist is completely stable in dimensions, and dimensional changes due to wetting and drying are basically cycling about a constant value.

"Growth" or continuing increase in length, leading to failure as a column, is not experienced as with some other materials.

12.3.5   Rubber Ring Joints

The important features of a rubber ring joint include

- It must provide flexibility and allow the design deflection at the joint, without leakage.
- It must transmit shear across the joint without allowing the consequent settlement to cause leakage.
- Jointing forces must not be excessive.
- Rubber ring interface pressures, throughout the life of the line, must be sufficient to prevent roof penetration into the line.
- For water supply and sewerage line it must be 100% effective in preventing infiltration.
- Joint surfaces, spigot and socket, must be manufactured with the necessary low tolerances.

Rubber ring joint in reinforced concrete pipe satisfy all these criteria; in fact the joint has been proven as capable of performing as design deflections to pressures well in excess of the capability of the parent concrete pipe.

12.3.6   Susceptibility to Corrosion

For normal installations, no protective treatment, coating or lining is required to ensure satisfactory service. Thus the problems with metallic pipes of ensuring continuity of the protective coating do not arise.

12.4   CONDITIONS REQUIRING SPECIAL CONSIDERATION

Reinforced concrete pipes, like most construction materials, have some service limitations and it is essential to recognize the conditions where special precautions or protective measures must be applied. The most important of these situations are outlined below:

12.4.1   Above Ground Exposure

Concrete pipes are designed for service underground and careful consideration is necessary if above ground or part buried installation are planned. Shrinkage and thermal stresses must be assessed – temperature differed of 30°C can arise in sun/shade situation with consequent cracking. Non-uniform shrinkages, restraints against thermal expansions and hot loe flow discharges may cause trouble.

Full submergence in water is satisfactory but partial immersion in water or soil should be avoided or allowed for in the design. The "wind and water zone" of a marine exposure comes into this category. Provision of a "sunshade" or cover of some form may be acceptable in some climatic conditions.

12.4.2 H₂S Attack

When certain combinations of circumstances occur in a sewer, pipes can be subjected to acid corrosion known as H_2S attack. Sulphate compounds existing in the sewage have a dissolved hydrogen sulphide (H_2S) fraction which can be released into the atmosphere. In aerobic conditions this H_2S is oxidized on the pipe surface by bacteria action with progressive reductions of pH and formation of sulphuric acid which attacks the cement compounds (in concrete or asbestos cement pipes). This process is only likely to occur with stale sewage (high biochemical oxygen demand), high temperatures, poor ventilation, flat grades and low flows, and where industrial wastes or other sulphate sources are admitted to the system. In general, aggressive combination of these adverse circumstances occurs only in the lager collecting trunk and outfall sewers.

In many cases the risk of H_2S attack can be minimized by proper sewer design, but in some circumstances it is economic to allow the H_2S to form and to protect the concrete against corrosion. Methods of protection used in Australia include.

- Sacrificial layers – additional concrete is provided in the pipe bore, to allow some internal concrete attack and still leave the pipe structurally sound.
- Use of cement with some aggregate – the concrete is made from acid soluble aggregate, ensuring that the acid is neutralized by the whole concrete matrix, thus significantly slowing the rate of attack.
- Use of internal linings to prevent access of the sulphuric acid to the concrete surface. Both liquid applied and performed plastic sheet materials have been used.

12.4.3 Aggressive Groundwaters

In some areas groundwaters can contain sulphates, chlorides, organic and inorganic acids or carbon dioxide, materials which can be aggressive to the cement compounds. Pure water can also be destructive.

With the exception of sulphates, when the expansive sulphoaluminate reaction described earlier occurs, the vehicle for attack is the free lime present in the concrete – this will react with acids, or leach out in pure soft waters. The initial removal of the lime then upsets the stability of the calcium silicate hydrates, resulting in the release of further lime, and also the corrosion proceeds.

The seriousness of groundwater attack is dependant on
- The concentration of the aggressive material.
- The permeability of the concrete.
- The rate of replenishment of the aggressive.
- The presence of other salts modifying the chemical reactions
- The cement type used.

Whilst the complete assessment of probability of attack is a job for an expert, it can generally be expected that.
- Sulphate attack will not be problem if Type D cement is used.
- Weak acids down to pH 5 have no effect whilst depending on type, concentration and replenishment rate, pH down to 3.5 can be innocuous.
- Pure water leaching is not a serious problem in most parts of Australia.
- Aggressive CO_2 higher than 20–70 ppm warrant consideration of protective measures.

Groundwater attack is usually best controlled by reducing rates of replenishment at the concrete surface be measures such as the use of impervious backfill, alkaline backfill, plactic wrapping of the pipe or application of an external organic coating.

12.4.4 Aggressive Effluents

The possibilities of internal attack by aggressive trade wastes are basically similar to those arising from aggressive groundwaters but because theses wastes are usually sufficient in quantity to maintain corrosion rates, acceptable concentration are lower than for groundwaters – for example, the pH limit is nearer 6 than 5. Protective measures similar to those adopted for H_2S attack are applicable for acid industrial wastes.

12.4.5 Reinforcement Corrosion

Because of the protection provided by the high pH of the concrete, attack on the reinforcement as distinct from the concrete is unusual. The only troublesome source can be very high chloride concentration e.g., brine. Steel corrosion due to pH reduction by carbonation, sometime experienced in cast in situ concrete is not a problem with precast pipe because the concrete impermeability prevents carbonation to significant depth.

12.5 CONCLUSION

Reinforced concrete pipe, as manufactured in Australia, is recognized word wide as the top quality and has a proven excellent service history in Australia. There are many situations where it is technically superior to other types pipe and conversely there are some application where special protective measures are required to ensure long life.

Indian experience is more or less similar, The first concrete pipe by spinning process was made in India by HUMES LTD in 1921 at Jamshedpur. They installed three factories in India at Jamshedpur, Kanan (near Nagpur) at Karari (near Jhansi) unfortunately they were not successfully. They give collaboration to the Indian Hume Pipe Co. Ltd. Is 1926. Since then, over 80 years; thousand of kilometers of concrete pipes are installed in India, which is gives satisfactory services.

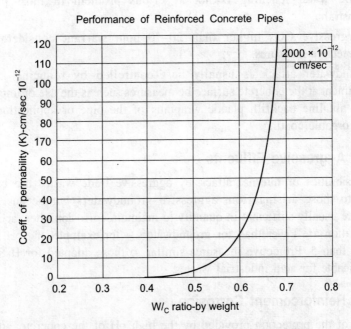

Performance of Reinforced Concrete Pipes

Fig. 12.1 Effect of initial W/c ratio on permeability

References

A full range of reference material can be obtained from the CPAA.A brief list of
   suggested reading is given below.

Chemistry of Cement

1. Lea and Desch – "The chemistry of cement and concrete".
2. W. Czernin – "Cement Chemistry and Physics for civil Engineers" (Crosby Lockwood & Sons Pvt. Ltd.).

Sulphate Resistant Cements

A.C.I. Monograph No 41968 "Durability of concrete construction".

Miller and Manson. Univ. of Minnesota Tech. Bulletin 184 (1951).

"Long-time Tests of concrete and Mortars Exposed to sulphate waters".

Permeability

1. P/I. Mahaffey – "Durable Concrete" Proceedings of CPAA National seminar 1972.
2. T.C. Powers, J.C. Hayes and H.M. Mann – "Permeability of Portland cement past" A.C.I. Journal Nov. 1954.

Pipe in clay Soils

SR and WSC Commission "Field study of Bedding of Asbestos cement Pipes in Dooen clays".

Seawater

B. Mather – "Effects of seawater on concrete" Highway Research Record No 113.

H_2S Attack

"Control of sulphides in Sewerage Systems" Edidted By P.K.B. Thistlethwayte (Butterworth 1972).

Aggressive Groundwaters

Biczok – "Concrete corrosion and concrete protection" (Akadamiai Kiado 1964).

CHAPTER 13

Maintenance of Concrete Pipelines

13.1 PREAMBLE

During normal operations, a concrete pipeline will give a satisfactory and trouble free service. However, accident and unforeseen circumstances may compel the maintenance authorities to undertake remedial measures. In this note, such situations which may cause trouble are indicated.

The main parts of concrete pipeline which may be affected by external influences, which are not anticipated are

1. **Damage to concrete due to corrosion**
2. **Joint problems**

If these two items are properly attended to, in design and execution, usually there should not be any trouble in the functioning of prestressed concrete pipeline.

In the subsequently paragraphs these factors are discussed in detail.

13.2 CORROSION OF SOIL

Use of synthetic manures or such other materials in course of time in the soil in which the pipeline or concrete structure construction is laid. In such cases, the soil conditions may change and becomes corrosive and may cause deterioration of coating.

The soil is considered corrosive when

(a) Exchangeable soil acids are equivalent to 50 ml of N/10 NaOH per 100 gms of dry soil.

(b) Chlorides as calcium chlorides : 0.1%

(c) Sulphates as SO_3 : 0.5%

(d) Magnesia as MgO.

13.2.1 Soil Resistivity

This gives an indication of the degree of corrosiveness. The following table gives the corrosion probability for various valued of soil resistivity for any concrete structure containing steel reinforcement.

Table 13.1   Resistivity and probability of corrosion

Resistivity: Ohms/cm³	Corrosion Probability
0 to 999	Severely corrosive
1000 to 4999	Corrosive
5000 to 10000	Low
Over 10000	Very low

In order to avoid corrosion probability, it is prudent to undertake inspection of the pipeline and the chemical analysis of the surrounding soil once in 12 months.

13.2.2 Effect of Temperature

As far as possible, prestressed concrete pipeline should not be laid above the ground. Shrinkage and thermal stresses, when temperature difference is around 30° may cause cracking. Non-uniform shrinkage, restraints against thermal expansion, and hot low discharge may cause trouble. Full submergence in water is satisfactory while partial immersion in water or soil should be discouraged. Exposure to marine condition should be avoided or a cover of some form may be provided. Contrary to normal belief that concrete pipe will not float, the pipe is uplifted if empty and submerged in water.

13.3 JOINT PROBLEMS

These are basically due to deflective civil engineering design of pipeline construction. Failure of a joint and subsequent leakage is mainly due to displaced rubber rings.

The main causes of this have been:
(a) Incorrect dimensions of joints.
(b) Incorrect laying practice.
(c) Insufficient anchor block size at bends.
(d) Insufficient stiffness of fittings.
(e) Roots of trees entering the annular gap and cracking the socket.

13.3.1

Incorrect dimensions of spigot may be due to defective mould. Daily opening and closing of the mould causes ovality in the spigot. It is necessary to rectify it from time to time. This is revealed, if the dimensions of every spigot is checked. Socket surface has to be finished smoothly to allow rolling of ring, at the same time its dimension must be accurate to achieve desired compression in ring to make joints water tight.

13.3.2

Use of feeler gauges and recording the travel of ring from socket face and inside clearance ensure proper jointing and positioning of ring. When the pipe is jacked or screwed, the dimension at three 60° position as shown below shall be taken and after the screw is released dimension at same position shall be repeated, if the difference is not appreciable then jointing is correct.

While negotiating curves the defection at each pipe joints should be uniform and not more than 2° In other words, a 10° curve is to be negotiated, this should be accomplished in 20 pipes. Some times there a twist in rubber ring when it is fitted on spigot. A twist in the rubber ring can be corrected by pulling it out, and releasing, if the ring remains in position there no twist.

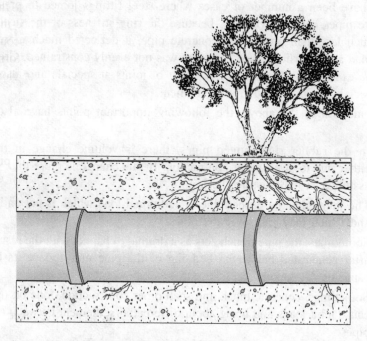

Photograph 17   Roots of trees entering the annular space between socket and spigot and exerting pressure and cracking

13.4 INSUFFICIENT ANCHOR BLOCK SIZE

In some cohesive soils, (e.g. black cotton soil) continuous stress at varying soil humidity causes creep and this tends to open up the rubber ring joints adjacent to the bend. It is therefore important in cohesive soils that thrust blocks be designed conservatively and joints in adjacent pipes are laid with little or no deflection. Block at special fitting has also to be executed properly.

When pipe is embedded in the wall of valve chamber, its settlement will be more than the portion of pipe which is outside. This uneven support may cause cracking hence needs proper bedding.

The pipes on pillars should be either supported at centre of gravity of pipe or at joint. Pillars, much away from centre of gravity of pipe exert pressure on the joint and cause leakage. On steep slopes, anchors may be necessary for few or for all pipes depending upon the degree of slope to avoid slipping and causing leakage.

Sluice or Butterfly valves must be anchored because when they are closed, thrust will be exerted on the valve.

13.5 INSUFFICIENT STIFFNESS OF FITTINGS

There have been a number of cases where steel fittings joined to prestressed concrete pipes have shown leaks because the ring stiffness of the steel fitting was much less than prestressed concrete pipe, it deflected much more under earth load and the rubber ring therefore was not evenly constrained. Great care must be exercised, to prevent embedment of joints at specials and also to see that specials are prevented from settlement.

In addition to the above, the following important points have also to be noted.

1. In the rubber ring jointed pipes, there is volume change in the soil, either due to swelling or contraction. If this change is more than what is assumed, it may disturb the joint.

2. The joints may be pushed due to additional load which is not anticipated and which may pass over the pipes.

3. Sometimes, the pipe trench acts as a drainage channel with the result, that, after penetrated below the bedding of the pipe may reduce its bearing capacity.

4. Sometimes, concrete block and such other construction done on the pipe, cause joint loads and bring about differential settlement between adjacent pipes.

5. Excavation behind anchor blocks within a particular zone is also dangerous. The thrust blocks may yield or move. Such a case, may arise

if a duplicate pipeline is laid just adjacent to the old pipeline. Any electric pole embedded in close proximity to pipeline may set in electric current in presence of water and may affect the wire.

6. When trees like banyan are close to underground pipeline (see Photograph 13.1) roots of trees enter the annular space between socket and spigot. If slight leak is there in the joint, the roots becomes thicker and exert pressure on the socket from inside and socket cracks, leakage is further increases and damage the joints. Such locations shown be observed by excavating the fill and joints inspected. For service connections it is generally desirable to provide a small pipe connected and parallel to main large than 400 mm size. This will ensure minimum interference of main supply, due to local operation and will protect the feeder main from unnecessary disturbance.

7. It may be said that the unsatisfactory performance of the pipes will be mainly due to inadequate precautions taken or incorrect data provided for design, or defective design itself. The pipe line which is already tested in field for anticipated maximum pressure, should not give away. It is only when the conditions which are not assumed initially or are changed, it may cause trouble.

13.6 OTHER CAUSES

Advances in concrete technology have made it possible to produce concrete of high durability without the need for low w/c ratio and high strength. The Table and Drawing reproduced below (by courtesy of TEL) identify the mechanism and causes of deterioration, and some of the principal factors that influence them.

Table 13.2 The nature or mechanism of six important forms of concrete deterioration and some of the controllable factors that influences these mechanism

Mechanism	Influencing Factors		
Sulphate	Cement Type	Cement	Content
ASR	Aggregate	Cement Type	Cement Content
Freeze/Thaw	Air Void System	W/C	
Carbonation	Grade	CPF	Curing
Chloride ingress	Cement Type	W/C	
Abrasion	Aggregate Type	Curing	Grade

Cross section of concrete showing, in schematic form, the following common causes of deterioration due to external chemical attack (due to sulphate and acid), internal chemical attack (due to alkali aggregate reaction, or delayed ettringite formation), external physical attack (including freeze/thaw, salt, weathering

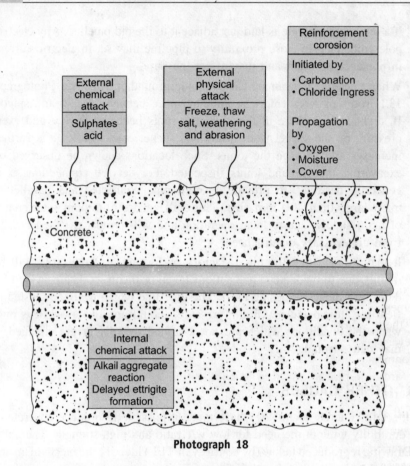

Photograph 18

and abrasion), and the corrosion of reinforcement (initiated by carbonation or chloride ingress, and propagated by oxygen, moisture and cover).

13.7 PIPE WALL HYDROKINETICS

As shown in Fig. 13.1, Example 1, with water at equal pressure on both side of a pipe wall, the concrete becomes saturated, stability is reached, and no water movement takes place. In Example 2, there is a differential pressure and regardless of magnitude, a hydraulic gradient causes movement of water through the wall, along with whatever salts, alkalis, sulfates, and other chemicals are in solution in the water. Even though this movement may be imperceptibly slow, it will provide continuous replenishment of any chemicals that may be in solution. Direction of flow is highly significant. If the aggressive water were on the right of the wall, the movement of non-aggressive water through the wall would tend to mitigate any effect. In either case, with no exposure to the atmosphere, there is no concentration effect.

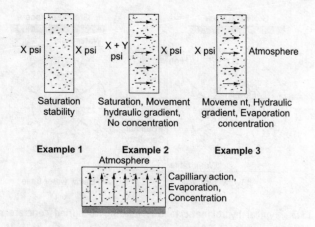

Fig. 13.1 Typical hydrostatic and hydrokinetic relationships

Examples 3 and 4 have an evaporative surface condition. The water movement is due to either a hydraulic gradient or capillary action. In either case, there would be a concentration at or near the evaporation surface of whatever chemicals are in solution. These considerations are not relevant to acid environments, since acid attack is essentially confined to the exposed surface. They are significant, however, in evaluating severity of sulfate or chloride exposures.

Full atmospheric exposure, as illustrated in Fig. 13.2, Example 1, can be a severe condition for concrete pipe. Depending upon climate and location, the exterior of the pipe could be subjected to freeze-thaw cycles, to thermal stresses, to chlorides in coastal areas, and to concentration effects of whatever salts or sulfates are in solution in the effluent.

Partial burial or immersion, as illustrated in Fig. 11.2, Example 2, can be a similarly severe exposure condition. Only a partially evaporative surface is provided, but the concentration effects are more complex since the source of

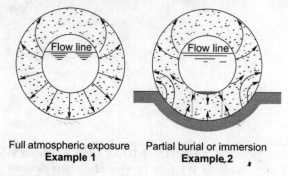

Full atmospheric exposure
Example 1

Partial burial or immersion
Example 2

Fig. 13.2 Typical hydrokinetic relating to atmospheric exposures of concrete pipe

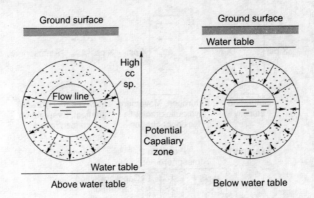

Fig. 13.3 Typical hydrokinetic factors relating to buried concrete pipe

salts or sulfates may be either the effluent or moisture from the ground entering the pipe wall through capillary action and moving toward the evaporative surface.

Buried pipe usually is not exposed to freeze-thaw or thermal stresses, and concentration effects are negligible. When installed above the water.

* **LANSAS PRODUCTS. – 1320 S SACRAMENTO ST, LODI, California 95240**
* Perkins P.H. "Repair, protection and water proofing of concrete Structure E and FN SPON – LONDON

Salient Feature of One of the Biggest Most Modern Plant for Manufacturing Concrete Pipes by Centrifugal Spinning Process

14.1 PURPOSE

The plant was specially constructed for manufacturing concrete pipe for storm water drainage of Baghdad in salan area in the republic of Iraq. For manufactures it will give an idea of the extent to which the technology has advanced.

14.2 DETAILS OF PIPES TO BE MANUFACTURED

Following are the details of pipes to be manufactured in 18 months. Some of these heavy duty pipes were provided with PVC sheet lining. The lining low to be provided immediately after final spinning by Hume process some pipes wire to be laid by jacking as shown in Table 14.1 below.

Table 14.1 Details of pipes: including external load

Dia mm	Tiickness mm	Extermal !oad kg/m	Normal Pipe Nos.	Jacking Pipe Nos.	Total Wt. Tons	PVC Lining	Wt. of Pipe kg
400	60	1900	216	–	121	–	560
500	70	2600	131	–	107	–	820
600	75	3000	1074	–	1117	–	1040
700	85	3350	86	–	119	–	1380
800	95	3750	2147	–	3822	–	1780
1000	115	5200	1809	62	4364	1103.	2330
1100	125	5600	320	–	890	539	2780

1200	130	6000	1183	–	3715	32	3140
1400	140	6700	131	–	513	594	3920
1500	150	7450	624	154	3509	–	4500
1600	165	8200	2309	135	12951	687	5290
1800	180	8950	1298	–	8393	560	6470
2000	190	9700	3290	359	27893	3131	7560
2250	215	11200	2611	899	34053	835	9630
2500	240	12700	182	333	6194	923	11950
			17447	1942	107766	8424	

Total no of pipes to be made   - 19389 Nos i.e., 4.84 km

Total weight   - 1,10,000 MT.

Normal pipes   - 17447 Nos

Jacking pipes   - 1942 Nos

PVC lining pipe   - 8424 Nos

14.3 MAN POWER PLAN

Estimated man power to be necessary for operating Hume pipe production plant.

Sl No	Description	Technician	Worker	Total
1	Manager Accountant	1		1
2	Foreman	3		3
3	Quality Control	1	1	2
4	Aggregate Handling		1	1
5	Batching Plant	1	1	2
6	Wire Straightening Machine		2	2
7	Steel Cage Production Equipment	7	7	14
8	Mould Assembly	6	12	18
9	Pipe Casting Operation	6	6	12
10	Overhead Crane Operation	3		3
11	Steam Curing		3	3
12	Pipe Inspection	1	2	3
13	Marking, Repair	1	4	5
14	Forklift Operation	4	4	8
15	Boilerman	1		1
16	Welder and Steel Fixer	1		1

17	Electrician	1		1
18	Storekeeper	1		1
19	Cook for your Worker		3	3
20	Kitchen Helper for us		2	2
21	Officer	1		1
22	Guard man	1		1
23	Others		5	5
	Total	**40**	**53**	**93**

Description

1 Manager

18 Storekeeper

Total   19

For other devices

Description	Technician	worker	Total
Cooks for worker and staff	40	53	93

Kitchen helper of Japan shaft

For production of 107,766 M.T. Pipe no of workers required 80 Nos.

Concrete Pipe productions per worker

$$\frac{80 \times 8 \times 18 \times 25}{107766} = \frac{288000}{107766} = 2.67 \text{ MT of heavy duly pipe.}$$

14.4 PRODUCTION PLANT

Layout and the plant was design by Neppon Hence pipe including manufacture and erection of the plant.

The work was started in October 1980 are completed in may 1982

Hence pipe manufactures shop low the follows.

(a) Area 42.0 m × 75 m.

(b) Layout

These production lines with following machines were provided.

(i) Ultra large size line for pipe Dia. from 2000 m to 2500 m.

(ii) Large size line for pipe Dia. from 1200 m to 2000 m.

(iii) Medium size line from pipe Dia. from 400 m to 1200 m.

Each time has following, daily production capacity for 8 hour- shaft.

(i) Ultra large size line   8 pipe per shaft

(ii) Large size line   15 pipe per shaft

(iii) Medium size line   18 pipe per shaft

Bar chart for production is given in Appendix A

14.4.1 Description

This plant is designed for production of reinforced spun concrete pipes and has approximately 6000 tons/month of productive capacity.

The Hume pipe production shop is provided with three production lines to be separated according to the pipe diameter.

These lines are ultra large size line (No. 1 line) for the pipe diameter from 2000 mm to 2500 mm, large size line (No. 2 line) for the pipe diameter from 1400 mm to 2000 mm and medium size line (No. 3 line) for the pipe diameter from 400 mm to 1200 mm.

14.4.2 Production Capacity

This plant has generally the production capacity as show in Table 14.2 – 15 per actual 8 hours daily work.

Table 14.2

Production line	Pipe diameter (mm)	Q'ty per 1 cycle time (pcs)	Daily capacity	
			Cycles	Q'ty (pcs)
Ultra large (No. 1)	2000-2500	2	4	8
Large (No. 2)	1400-2000	3	5	15
Medium (No. 3)	400-1200	3	6	18

14.5 ARRANGEMENT OF PIPE CASTING EQUIPMENT

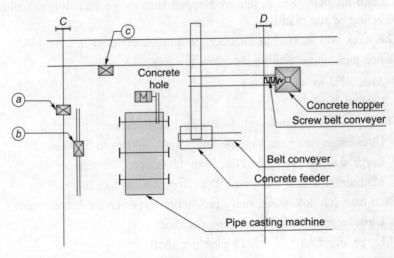

Fig. 14.1

For Operator

(a) Control Panel of Concrete feeder

(b) Control Panel of Pipe Casting Machine and Concrete Feeder

(c) Control Panel of Pipe Casting Machine

14.6 PLANT FACILITIES

This plant basically consists of the following facilities:

(1) Hume pipe production shop.

(2) Batcher plant with concrete materials stock facilities. Two (2) sets of batcher plants are provided, one for Hume pipe shop and the other for general use.

(3) Electricity supply facilities

(4) Water supply facilities

(5) Steam supply facilities

(6) Water curing yard of produced Hume pipe.

(7) Stock yard of produced Hume pipe.

(8) Site office.

(9) Accommodation facilities.

(10) Testing facilities.

(11) Warehouse.

14.7 MACHINERY

Main machineries for production of Hume pipe:

Name	Q'ty	Description
No. 1 Casting machine (ultra large size)	1	75 kW –5 00~2000 RPM 2 beds
No. 2 Casting machine (large size)	1	75 kW –500~2000 RPM 3 beds
No. 3 Casting machine (medium size)	1	55 kW 3 beds
No. 1 Overhead crane	1	20 ton 13 m span
No. 2 Overhead crane	1	15 ton 13 m span
No. 3 Overhead crane	1	6.5 ton 13 m span
Batcher plant (EMC–55)	1	55 cup. m/hour output
Concrete carrier	1	2 cup. m – 45 m/min.
No. 1 Concrete feeder	1	2 cup. m – 50 m/min.
No. 2 Concrete feeder	1	2 cup. m – 50 m/min.
No. 3 Concrete feeder	1	2 cup. m – 50 m/min.
No. 1 Mould assembly equipment		Electrical chain block 10 ton × 2, 2 ton × 1

No. 2 Mould assembly equipment		Electrical chain block 2 ton × 3, 5 ton × 1
No. 3 Mould assembly equipment		Electrical chain block 1 ton × 2, 2.8 ton × 2 2 ton × 1
No. 1 Steel cage production machine	1	180 KVA Ø 1350 ~ Ø 2600 mm
No. 2 Steel cage production machine	1	87 KVA Ø 900 ~ Ø 2000 mm
No. 3 Steel cage production machine	1	31.3 KVA Ø 400 ~ Ø 1200 mm
Wire straightening machine	1	Ø 3 ~ Ø 7 mm Cutting length 3 m
Boiler	1	3 ton/hour Max. pressure 10 kg/cm^2
Compressor	1	Pressure 7 kg /cm^2, 7 m^3/min. 37 kW
External load testing machine	1	Man. 50 ton Stroke 3800mm
Hydraulic pressure testing machine	1	Hydraulic pump Blind flange etc.
Water feed pump	3	11 kW – 0.72 m^3/min., head 35 m 11 kW – 0.83 m^3/min., head 30 m 11 kW – 1.0 m^3/min., head 30 m
Discharge pump	2	2.2 kW – 0.5 m^3/min., head 14 m 3.7 kW – 0.7 m^3/min., head 14 m
Folk lift	4	15 ton × 1, 13.5 ton × 2, 2 ton × 1
Shovel loader	1	0.9 cub. M

14.8   SPECIFICATIONS FOR REINFORCED SPUN CONCRETE PIPE FOR STORM SEWER DRAINAGE SCHEME FOR THE CITY OF BAGHDAD

14.8.1 Scope

These specification cover reinforced spun concrete pipe intended to be used for STORM SEWER DRAINAGE SCHEME FOR THE CITY OF BAGHDAD.

14.8.2 Classes

Pipe manufactured according to these specifications shall be classified, as show in Table 14.3, into common pipe for open-cut laying method and pressure pipe for rising main pipe line according to use, and further into type B, C and J according to shape.

Table 14.3

Classes	Type – J	Type – B	Type – C
Common Pipe	—	400 ~ 700	800 ~ 2500
Jacking Pipe	600 ~ 2500	—	—
Rising Main	1000 ~ 2500	—	—

Type-B

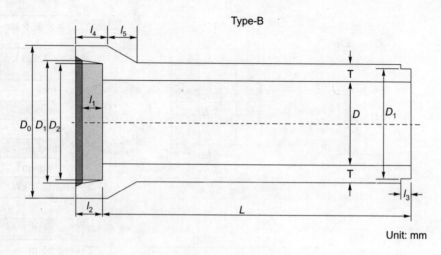

Unit: mm

Fig. 14.2 Type-B

Table 14.4 Type – B

Dia.	D_1	D_2	D_3	D_4	T	l_1	l_2	l_3	l_4	l_5
400	528	524	510	640	60	70	95	36	125	110
500	648	644	630	780	70	70	95	36	130	125
600	758	754	740	900	75	75	100	36	135	145
700	878	874	856	1040	85	75	105	40	140	160

** Effective length: $L = 2430$ mm

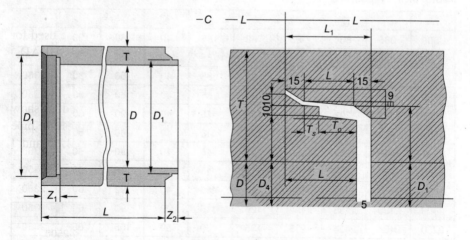

Fig. 14.3 Type C

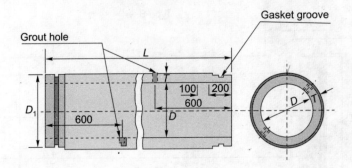

Fig. 14.4

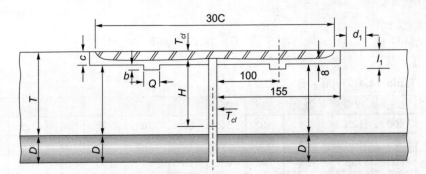

Fig. 14.5

Table 14.5 Types – C

Dia.	D_1	D_2	T	l_1	l_2	l_3	l_4	l_5	L
800	880	864	95	90	85	40	60	30	2360
900	998	974	110	90	85	40	60	30	2360
1000	1108	1084	115	90	85	40	60	30	2360
1100	1218	1194	125	90	85	40	60	30	2360
1200	1322	1298	130	120	115	40	90	60	2360
1400	1532	1508	140	120	115	40	90	60	2360
1500	1642	1618	150	120	115	40	90	60	2360
1600	1756	1732	165	120	115	40	90	60	2360
1800	1970	1946	180	120	115	40	90	60	2360
2000	2180	2156	190	120	115	40	90	60	2360
2250	2454	2424	215	135	130	50	105	65	2360
2500	2728	2698	240	135	130	50	105	65	2360

Tolerance on Dimensions

4. Tolerance on dimensions shall conform to Table 14.6

Table 14.6 Type – J

Dia.	D	T	D_1	L
600 ~ 1200	±6	+6 −3	±3	
1400 ~ 1600	±8	+8 −4	±4	+10 −5
2000 ~ 2250	±10	+10 −5		
2500	±12	+12 −6	±5	

Table 14.7 Type – B

Dia.	D	$D_1 D_2 D_3$	D_4	T	l_1	l_2	l_3	l_4	l_5	L
2000 ~ 2250	±4	+3 −2	+10 −5	+4 −5	+5	+5	+5	+10 −5	+10 −5	+10 −5

14.9 PIPE REINFORCEMENT DETAILS

Table 14.8

Pipe Dia. mm	Thickness mm	Circular Reinfocement			Longitudinals
		Wire pitch and Dia. mm	Steel area Mm^2/cm	As/Ac %	Number and Dia. Wt/cage
400	60	64 – 3.5 Ø	1.50	0.25	10 – 3.5 Ø
500	70	64 – 3.5 Ø	1.50	0.21	10 – 3.5 Ø
600	75	64 – 3.5 Ø	1.50	0.20	10 – 3.5 Ø
700	85	56 – 4.5 Ø	2.83	0.33	12 – 4.5 Ø
800	95	50 – 4.5 Ø	3.18	0.33	12 – 4.5 Ø
1000	115	50 – 4.5 Ø 62 – 4.5 Ø	3.18 2.56	0.50	12 + 12 – 4.5 Ø
1100	125	46 – 4.5 Ø 58 – 4.5 Ø	3.45 2.74	0.50	12 + 12 – 4.5 Ø
1200	130	41 – 4.5 Ø 53 – 4.5 Ø	3.87 2.99	0.53	12 + 12 – 4.5 Ø
1400	140	58 – 6.0 Ø 78 – 6.0 Ø	4.87 3.67	0.61	12 + 12 – 6.0 Ø
1500	150	53 – 6.0 Ø 70 – 6.0 Ø	5.33 4.04	0.62	12 + 12 – 6.0 Ø

1600	165	43 – 6.0 Ø 58 – 6.0 Ø	6.58 4.87	0.69	12 + 12 – 6.0 Ø
1800	180	38 – 6.0 Ø 51 – 6.0 Ø	7.44 5.54	0.72	12 + 12 – 6.0 Ø
2000	190	45 – 7.0 Ø 60 – 7.0 Ø	8.55 6.41	0.79	12 + 12 – 7.0 Ø
2250	210	46 – 8.0 Ø 62 – 8.0 Ø	10.93 8.11	0.89	12 + 12 – 7.0 Ø
2500	240	34 – 8.0 Ø 46 – 8.0 Ø	14.79 10.93	· 1.07	16 + 16 – 7.0 Ø

14.10 MANUFACTURE

14.10.1 Flow Process Chart

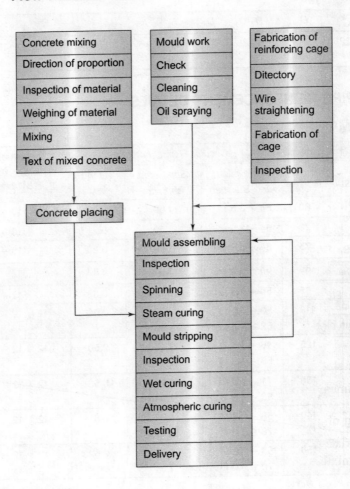

Fig. 14.6 RCC pipes (2500 mm Dia) with PVC lining

14.11 MANUFACTURE

14.11.1 Design Concrete Strength kg/cm^2

Compressive strength of concrete shall be tested for the standard column specimen with 10 cm diameter and 20 cm height, and the strength shall not fall the value show in Table 14.9.

Table 14.9 Concrete strength

Pipe Dia. class	400 ~ 700	800 ~ 1200	1400 ~ 1800	2000 ~ 2500
Common pipe and Jacking pipe	320	350	380	420
Rising main pipe	–	380	420	450

Maximum size of coarse aggregates

The maximum size of the coarse aggregate shall be 20 mm.

Measuring of materials

The materials of concrete shall be measured in weight provided that water and liquid admixtures may be measured by volume.

14.11.2 Concrete Mixing

Concrete shall normally be mixed in a mechanical mixer, and mixing shall be continued until there is a uniform distribution of the materials and the mass is uniform in color and consistency, but in no case shall be mixing be done for less than two (2) minutes. The concrete shall be placed before setting has commenced. It should be ensure that the concrete is not dropped freely so as to cause segregation.

14.11.3 Concrete Consolidation by Spinning

At the early time of spinning, rotation of the mould shall be as slow as placed concrete will distribution uniformly throughout the entire length of the mould. Time and speed of spinning shall be such that the pipe may be freely consolidated. Concrete of the additional layer shall be placed after the excess water, squeezed from the wall, has removed completely from the inner surface of the pipe. At the finish of spinning, obstacles such as projecting stone or rubbish shall be removed, and the dents shall be filled with mortar and then smoothed.

14.11.4 Curing

The pipe shall be cured by such method that produces satisfactory results. For conducting normal steam curing, it shall be, as a rule, carried out as follows:

(a) The pipe, as being contained in its mould, shall be placed on a curing concrete floor and covered with sheets, free from outside drafts, and the steam shall be introduced so that the temperature of in the hood may rise uniformly.

(b) The steam curing shall not be performed unless two (2) hours or more time have elapsed since mixing of the concrete.

(c) The temperature of the curing hood shall be raised at a rate not exceeding 20°C per hour to a maximum of 65°C.

(d) The pipe shall be removed from the curing place after the temperature has been gradually lowered until the difference with ambient temperature has become unsignificant. Standard steam curing time shall be such as show in Table 14.10.

Table 14.10

Pipe Dia. / Steaming Hour	400 ~ 1200	1400 ~ 2500
1	Nearly 60°C rise by three hours.	Nearly 60°C rise by three hours.
2		
3		

4	60°C	60°C
5	60°C	60°C
6	60°C	60°C
7	–	60°C

14.11.5 Assembling and/or Stripping of Mould

The pipe shall be taken out from the mould after the temperature of the pipe has been gradually lowered until the difference with ambient temperature has become unsignificant. Excess shock or impact shall not be given at the time of stripping. The inner surface of the mould shall be clean and dreg of concrete may not be permitted.

14.12 WORKING TIME SCHEDULE

Basically, workers shall be occupied to work nine (9) hours including one (1) hour rest per normal working day and to work twenty six (26) days per month. We are planning to take following working shift for each production line.

(1) Ultra large size line

Actual 8 hours work		9 Months
Actual 9.5 hours work		2 Months
Actual 11 hours work		9 Months
	Total	20 Months

(2) Large size line

Actual 8 hours work		11 Months
Actual 9.5 hours work		9 Months
	Total	20 Months

Medium size line

Actual 8 hours work	18 Months

14.12.1 Productivity by Spinning

Activity diameter	No 1 Ultra large 2000 to 2500	No 2 Large size 1400 to 2000
Feeding of concrete to steel mould	5 to 8.5 Minutes	6 to 8 Minutes
Spinning of mould and compaction of concrete and taking away water	25 to 34	17 to 21 Minutes
Feeding of concrete for second layer	5 to 8.5	5 to 8 Minutes

Spinning of mould rotating of steel for compacting of concrete	18 to 25	11 to 15 Minutes
Finishing for inner surface of pipe	12 to 13	8 to 11 Minutes
Lifting of mould and bringing new mould	15 to 20	21
TOTAL : Minutes	81 to 110 Minutes	69 to 84 Minutes

No of runns/Shaft		
No of pipes per cycle	2	3
Pipes per shaft		
No of runns per shaft	4	5
No of pipes	8	15

14.12.2 Cycle Time of Mould Work

	Work Item		Production Line		
	Description	Item No	No 1	No 2	No 3
Stripping	1. Detaching of end ring rods Cores (for Grouting and Tensioning)	(2)	1'30"	1'	1'
	2. Shifting of Mould on Stripping Support	(3), (4)	1'30"	1'	1'
	3. Detaching of Short-tie rods	(5)–a,b	4'	3'	2'30"
	4. Retreating of U-Fork	(5)–c	30"	30"	30"
	5. Fixing of Lifting Jigs for end ring	(5)–d	30"	30"	30"
	6. Detaching of tie rods (2 pcs.)	(5)–d	10"	10"	10"
	7. Detaching of end ring (by Hydraulic Jack)	(5)–e	1'30"	1'	1'
	8. Detaching of seam Bolts	(6)	20"	20"	20"
	9. Inserting of U-Fork	(6)	30"	30"	30"
	10. Lifting up of Topside Mould	(7)–a	30"	30"	30"
	11. Drawing out of Pipe	(7)–b	1'	1'	1'

Assembling	12. Placing down of Topside Mould shell	(1)	12"	11'30"	9
	13. Tightening of Seam Joint	(1)	3'30"	3'	2'30"
	14. Taping at the Parts of Casting Plate and end ring	(2)	4'30"	3'	2'30"
	Taping only at the Parts Of end rings		(1')	(1')	(1')
	15. Applying of Form oil	(3)	1'	30"	30"
	16. Inserting of cage	(4)	3'	1'	1'30"
	17. Tightening of tie rods	(5)–a,b,c, d,e	7'30"	1'	5'
	18. Fixing of Casting Core (for Grouting and Tensioning)	(6)–a,b	8'	5'	4'
	19. Carriage	(7)	5'	5'	5'
	Total		45"	34'30"	30'30"

Note: This table to be applied for production of Jacking pipe.
In case of production of common C and B type pipe, work item No. 1 and No 18 to be deleted.

Table 14.11 Abstract of time in minutes required for operation per pipe

Diameter	2500	1200	900
	1	2	3
a. Spanning mins.	110 mins	70/2 = 35	66/3 = 22 mins
b. Assembling and stripping	45 mins	34–30 mins. sec	30–30 mins. sec
c. Cage making	50 mins	20 mins	15 mins

Spinning time govern the productively

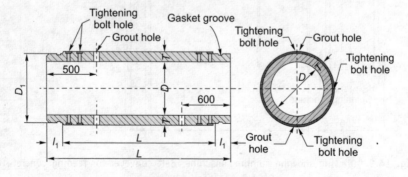

Fig. 14.7 Standard pipe

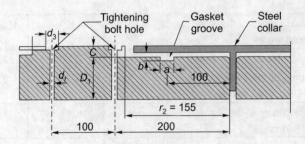

Fig. 14.8 Details of a standard pipe

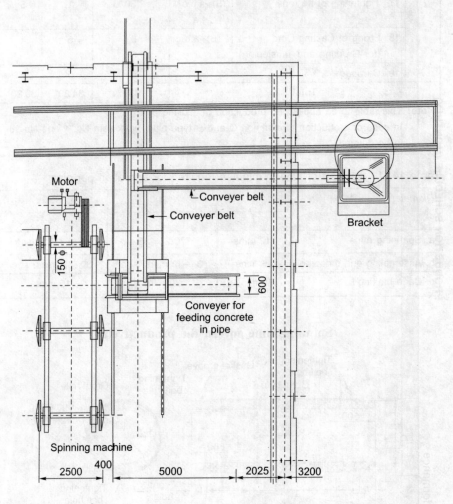

Fig. 14.9 Drawing showing spinning machine, Belt conveyer for feeding concrete in pipe for converying concrete

14.12 Performance

PRODUCTION SCHEDULE (MONTHLY)

UNIT:PSC

Dia. mm	Pipe type	1980			1981												1982						Total PSC	Unit Wt. (kg/pc)	Total Wt. (TON)
		10	11	12	1	2	3	4	5	6	7	8	9	10	11	12	1	2	3	4	5	6			
400	COMMON-B					65	65	65	33															228	560	128
500	COMMON-B	30	55	55	55																			195	820	160
600	COMMON-B			15	65	108	108	108	108	108	108	108	108	65	20	20	20	20	20	20	13			1142	1040	1188
700	COMMON-B			45	45																			90	1380	124
800	COMMON-C										75	175	175	175	175	153	275	269	270	261	276			2279	1780	4057
1000	COMMON-C					165	183	183	183	183	208	168	70	88	103	73				103	104			1814	2330	4227
1000	JACKING			13											5	5	3	3	3	3				35	2400	84
1000	DELIVERY																			5	25			30	2330	70
1100	COMMON-C															53	97	93	93					336	2780	934
1200	COMMON-C					163	160	160	160	160	60		97	93	93	43								1189	3140	3733
1400	COMMON-C					25	25	25	25	25	10													135	3920	529
1400	DELIVERY																									
1500	COMMON-C			100	100	100	100	100	100	25														625	4500	2813
1500	JACKING					25	25	10																60	4630	278
1500	DELIVERY															5	25	25	25	25				105	4500	473
1600	COMMON-C	150	150	149	149	149	149	124	149	124	124	124	124	124	124	124	120	121	30	30	60			2398	5290	12685
1600	JACKING					11		8						25	15	6	25	21	26					137	5450	747
1800	COMMON-C	101	101	101	101	101	101	144	109	82	70	41							100	50	145			1347	6470	8715
2000	COMMON-C									160	168	211	269	269	269	298	312	292	281	310	140			2979	7560	22521
2000	JACKING				25	25	25	25	45	25	38	45	35	23	23	25	25	25	25	25				459	7780	3571
2000	DELIVERY																									
2250	COMMON-C	132	123	213	223	223	223	248	248	223	248	223	123	123	56	19								2648	9630	25500
2250	JACKING	25	23	23	44	43	43	48	48	77	48	48	52	56	48	48	48	48	48	48	48			914	9910	9058

Contd...

250C	COMMON-C	32	32	32	32	32	32															192	11950	2294
2500	JACKING	25	25	25								18	18									111	12300	1365
2000	RISING							6	5	5	5	32	30	30	30	60	60					263	7560	1988
TOTAL NOS	(PCS)	465	454	716	784	1145	1199	1198	1185	1192	1163	1148	1058	1076	986	972	1008	1012	1016	1005	929			
TOTAL WT	(TON)	3656	3549	4970	5182	6209	6433	6394	6343	6753	6542	6343	5591	5683	4887	5060	4967	4773	4885	4843	4179			

PRODUCTION SCHEDULE (DILY AVERAGE)

Dia. (mm)	Pipe Type	1980				1981												1982				REMARKS MOLD	REMARKS MOLD(PSC) CASTING
	Type	9	10	11	12	1	2	3	4	5	6	7	8	9	10	11	12	1	2	3	4	MOLD	MOLD(PSC) CASTING
	(col #)	1	2	3	4	5	6	7	8	9	10	11	12	13	14	15	16	17	18	19	20		
ULTRA LARGE																							
2500	COMMON=C		1.0				0.52	1.0	1.0	1.0	1.0	1.0	0.76									2	
2500	JACKING	1.0		1.0	1.0	1.0	0.48															1	
2500	DELIVERY													1.0	1.0	1.0	1.0	1.0	2.0	1.84			2
2250	COMMON-C	4.0	4.0	4.0	7.0	7.0	7.0	7.0	7.0	7.0	7.0	5.0	5.0	5.0	5.0	5.0	5.0	5.4	6.0	2.0	0.04	7	
2250	JACKING	3.0	3.0	3.0	2.0	2.0	2.0	2.0	2.0	2.0	2.0	2.0	2.0	2.0	2.0	2.0	2.0	0.96				4	
	TOTAL	8.0	8.0	8.0	10.0	10.0	10	10.0	10.0	10.0	10.0	8.0	8	8.0	8.0	8.0	8.0	7.36	8.0	3.8	0.04	14	2
LARGE																							
2000	COMMON-C	7.0	7.0	7.0	8.0	8.0	8.0	8.0	8.0	8.0	8.0	6.0	6.0	6.0	8.0	7.0	6.0	6.0	6.0	3.6		8	
2000	JACKING	1.0	1.0	1.0	1.0	1.0	1.0	0.36														1	
2000	DELIVERY															1.0	3.0	3.0	2.44				3
1800	COMMON-C	3.0	3.0	3.0	3.0	3.0	3.0	3.64	4.0	4.0	4.0	4.0	4.0	4.0	3.0	3.0	0.28	3.0				5	
1600	COMMON-C	4.0	4.0	4.0	6.0	6.0	6.0	5.0	5.0	5.0	5.0	5.0	5.0	5.0	5.0	5.0	6.0	6.0	5.36			6	
1600	JACKING							1.0	1.0	1.0	1.0	1.0	0.4									1	
1500	COMMON-C	2.0	2.0	2.0	3.0	3.0	3.0	3.0	3.0	3.0			0.6	0.36								3	
1500	JACKING	1.0	1.0	0.32										0.64								1	
1500	DELIVERY														1.0	1.0	1.0	0.2					1
1400	COMMON-C																	1.0	1.0	0.24		1	4
	Total	18.0	18.0	17.32	21.0	21.0	21.0	21.0	21.0	21.0	18.0	16.0	16.0	18.0	18.0	18.0	17.28	16.2	14.8	3.84		26	4

Contd...

MEDIUM	Type	1	2	3	4	5	6	7	8	9	10	11	12	13	14	15	16	17	18	19	Nos.
1200	COMMON-C	4.0	4.0	4.0	4.0	4.0	4.0	4.0	4.0	4.0	4.0		4.0								4
1100	COMMON-C			3.0	3.0	3.0	0.8					3.3									3
1000	COMMON-C	5.0	6.0	5.52	6.0	6.0	6.0	6.0	6.0	6.0	6.0	2.84	6.0	6.0							6
1000	JACKING	1.0		0.48																	1
800	COMMON-C	6.0	6.0	6.0	6.0	6.0	6.0	6.0	6.0	6.0	6.0	6.0		1.88	6.0						7
700	COMMON-B						2.2	1.24													2
600	COMMON-B	2.0	2.0	2.0	2.0	2.0	2.0	2.0	2.0	2.0		0.7	4.0	2.0		2.28	4.0	4.0			4
500	COMMON-B											0.8			0.44		2.0	2.0			2
400	COMMON-B											0.8				1.84	2.0	2.0			2
TOTAL		18.0	18.0	21.0	21.0	21.0	21.0	19.24	18.0	18.0	16.0	14.44	14.0	9.88	6.44	4.12					31
GRAND TOTAL		26.0	44.0	43.320	52.0	52.0	52.0	52.0	50.24	46.0	42.0	41.76	42.0	40.44	40.0	35.16	30.0	26.92	7.68	0.04	71

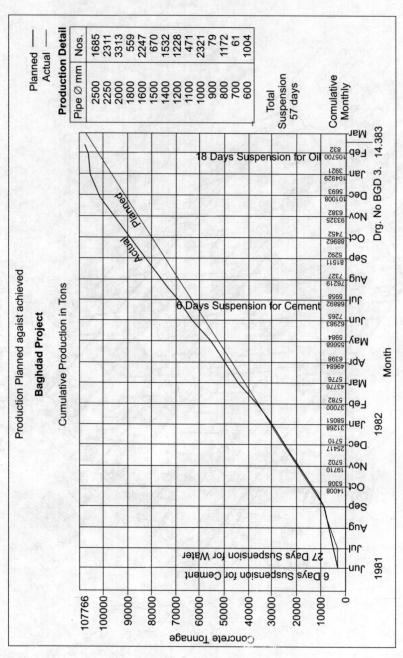

Fig. 14.10

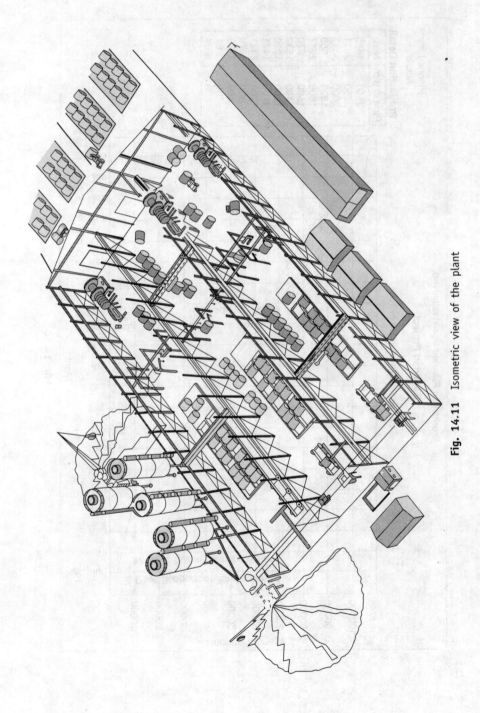

Fig. 14.11 Isometric view of the plant

Photograph 19 Isometric view of the plant at Baghdad

Photograph 20 RCC Pipes (2500 mm Diameter) with PVC lining

14.13 WORKING TIME SCHEDULE

Basically, workers shall be occupied to work nine (9) hours including one (1) hour rest per normal working day and to work twenty six (26) days per month. We are planning to take following working shift for each production line.

   (1) Ultra large size line

      Actual 8 hours work   9 months

      Actual 9.5 hours work   2 months

Actual 11 hours work 9 month

Total 20 months

(2) Large size line

Actual 8 hours work 11 months

Actual 9.5 hours work 9 months

Total 20 months

(3) Medium size line

Actual 8 hours work 18 months

Dia mm	Thickness mm	External Load Kg/M	Normal Pipe Nos.	Jacking Pipe Nos.	Total Wt. Nos.	PVC Lining Nos.
400	60	1900	216	–	121	–
500	70	2600	131	–	107	–
600	75	3000	1074	–	1117	–
700	85	3350	86	–	119	–
800	95	3750	2147	–	3822	–
1000	115	5200	1809	62	4364	1103
1100	125	5600	320	–	890	539
1200	130	6000	1183	–	3715	32
1400	140	6700	131	–	513	594
1500	150	7450	624	154	3509	–
1600	165	8200	2309	135	12951	687
1800	180	8950	1298	–	8398	560
2000	190	9700	3290	359	27893	3131
2250	215	11200	2611	899	34053	835
2500	240	12700	182	333	6194	923
			17411	1942	107766	8424

Illustration of the Process by Photographs
Special Arrangements of Every Operation are Worth Nothing

OBJECT

A process is better understood if photographs are provided. In addition, if short description of process is given, the intricacies are clear, Developments in production are aimed at increasing productivity, reducing labours without affecting quality, all These are achieved in the process given below. The crux is that the various parts of the mould (shell, end rings and tie rods) are not separated much, but allowing easy removal of pipe from the mould by using proper equipment.

14A SPINNING

14A.1 General Arrangement

The concrete for the pipe is fed by belt feeder. There is only one feeder for feeding even four pipes. The belt feeder has lateral movement, so that when feeding in one pipe is completed, it comes out and laterally shifted in front of another pipe and then it feeds another pipe.

Concrete is supplied to the feeder by means of cross belt feeder which is supplied concrete from the hopper. Such hoppers are fixed in a line across the machine and are fed by a trolley which carries concrete from the mixer of batching plant.

14A.1.1 Concrete feeding and spinning operation

All the controls of machine, belt feeder are with the moulder at one place who stands on opposite side of the pipe feeder belt.

When thickness of concrete is more than 70 mm, the concrete is fed in two layers so that entire water from the thickness of concrete is removed to the maximum extent.

When the compaction of first layer of concrete is going on, the water that comes out, is removed by water jet. When complete water is removed, concrete for second layer is fed. After proper leveling, it is smoothed by a steel patti and then by a sleeker. (Rubber sheet piece fixed to steel pipe).

After completing the spinning and smoothening, the mould is stopped; inside surface is checked and then the mould is lifted by overhead crane with lifting tackle and taken to the curing area and lowered.

14A.1.2 Lining with PVC sheet (Humes process)

For such pipes, spinning is stopped earlier, when the inside surface is soft. The mould is lifted and placed on pair of runners placed parallel to the main spinning machine. PVC lining sheet which has projecting strip outside, is then fixed inside the pipe and then the sheet is pressed by vibrating screed so that the projections are pressed inside concrete.

14A.2 Stripping

Mould from curing area with pipe inside, is lifted by over head crane and brought to stripping area and lowered on wooden battens.

Hair pin lever is then inserted in pipe, from one end. Pipe is slightly lifted and turned so that the hole for lifting tackle comes on top and seams naturally at horizontal working level. This is possible, because there are rollers on the hair pin lever. It is then removed.

End ring is then suspended from the lifting jig suspended from swivel arm. Remaining tie rods are then removed. The end ring is then pushed out by two jacks working on each side. The end ring which is now separated from mould shell is push out by swivel arm.

Seam bolt are then removed by wrenched. Upper end of mould shell is then lifted up by chain blocks (mould shell is in two parts).

Pipe is then taken out of mould by hair pin lever, Mould shell inside surface and seams are then cleaned.

14A.3 Assembly of the Mould

The upper half of the mould which was lifted up is lowered by chain blocks and the seam bolts are then fixed and tightened. A tape is fixed on the seam from inside which will prevent leakage of water.

Cage is then inserted (single or double) inside the mould by means of hair pin lever. Before that mould oil is sprayed uniformly by a spray. The hair pin lever is then removed.

End rings which are suspended form swivel arms are then brought close to the mould tie rods are then inserted and tightened with end rings in position in the sequence as shown. Starting from the bottom and then going to the top. Lifting jigs for end rings are then detached. The tie rods are tightened in such a way that 25 mm bolt remains out. The mould is then lifted by overhead cranes and taken to the assembled mould area. The tie rods are only 600 mm long.

Note: *Please observe the photographs very carefully so that the modifications made in the process can be appreciated.*

(a) Spinning Operation

Photograph 21A   Lifting assembled mould and transferring it to the spinning machine. Note the tackel for lifting

Photograph 22A   The assembled mould is brought on the spinning machine and lowered

Photograph 23A Concrete feeding is started from on end. After completing one pipe the belt feeder is shifted to another pipe and concrete is fed

Photograph 24A Compacting of concrete by spinning, of the first layer of concrete. Belt feeder is seen on left

Photograph 25A While spinning is going on for the first layer of concrete, water is removed by water jet

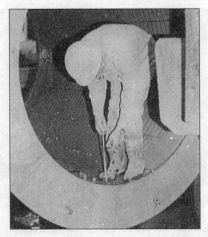

Photograph 26A After completing the spinning of the first layer, location of the reinforcement cage is checked on the machine it-self

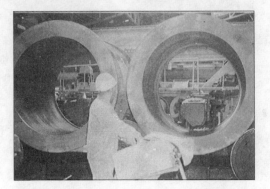

Photograph 27A Concrete feeding for secound layer one by one, with same belt feeder

Photograph 28A Spinning and compaction of concrete for secound layer. Rough finish by steel patti

Photograph 29A   Smooth finish to inside surface by steel patti

Photograph 30A   At the conclusion of spinning, water is removed by a sleeker, (Rubber sheet piece fixed to steel pipe)

Photograph 31A   Inside surface is checked before the mould is lifted up

Photograph 32A   Mould is then lifted and taken to curing area

Photograph 33A   PVC sheet is placed inside pipe such that projecting keys are in contact with concrete. After that a beam with vibrates is inserted inside. Before insertion of sheet pipe surface to be soft

Photograph 34A   Vibrators are operated and pipe rotates centrifugally. The vibration is done till the projecting keys are fully embedded

(b) Stripping Operation

Photograph 35B Shifting the mould from curing area

Photograph 36B Lowering the mould on stripping base, (without giving a shock,) by putting packing under the mould

Photograph 37B Insert hair pin lever in the mould and left it up a little. Turn the mould on the rollars on hair pin lever so that the hole for lifting pin come at the top. The mould seams well then come to working horizontal level

Photograph 38B   Lower the mould on strepping bench and start loosening tie rod bolts

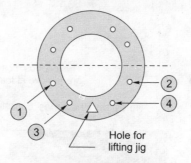

Order of losening tie rods

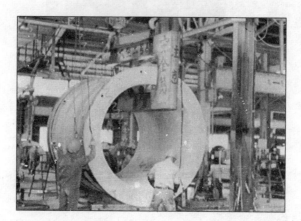

Photograph 39B   Remove the lower tie rods except 2 Nos

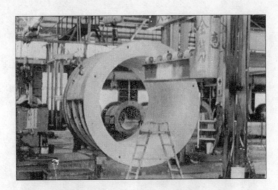

Photograph 40B   Remove hair pin lever. Not to remove 2 nos. of tie rods at bottom as shown below only loosen them to avoid end ring dropping down

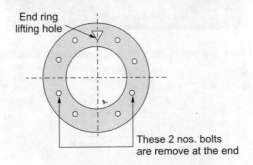

Holes on end rings

Photograph 41B   Fixing of lifting jigs in the hole as shown above for end ring. After fixing the lifting jigs at end rings remove two remaining tie rods

Photograph 42B   End ring is suspented from lefty jig at top. Remaining tie rods are then remove

Photograph 43B   Detaching of end rings by using two hydraulic jacks, two at a time. (from each side) By keeping even hydraulics pressure on jacks

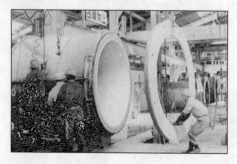

Photograph 44B   Remove end rings suspended on chain block fitted on swivel arm

(c) Assembling Operation

Photograph 45B Remove seam bolts by pneumatic or electric branch. See that on the other side, the bold has stopper

Photograph 46B Lift-up one side of top shell by chain block

Photo 47B Drawing out of pipe by hair pin lever and then lowering the top shell of mould

Photograph 48B   The top sheel os lowered and the seam bolts are tightened

Photograph 49B   Fix tape at seam, fix the tape at both seam joints

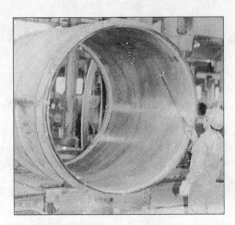

Photograph 50A   Apply mould oil uniformly by air spray. Density of oil to be selected
as per pipe diameter

Mould and
tape

Horizontal &
seam tape

Photograph 50B Tapeing, both at the part of end ring joint with shell and Tapeing at
the seam joint mould joints

Photograph 51C Inserted the cage taking care not to disturb the shape of cage, with
help of hair pin lever

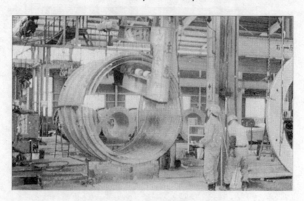

Photograph 52C Insert the cage with hair pin lever, taking care not to get the cage
out of shape

Photograph 53C Bring the suspended end ring near mould after cleaning the mateing face with mould

Photograph 54C Fixing of end ring. Fixing the lower tie rods first

Photograph 55C   Tightening of tie roads (lower side)

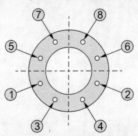

Order of tightening tie rods

Photograph 56C   Tightening tie rod

Photograph 57C   Tightening of the tie rod (Upper side)

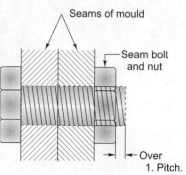

Seams of mould

Seam bolt and nut

Over 1. Pitch.

Photograph 58C   Detaching of lifting jigs from end ringing and check all tie rods tightened completely

CHAPTER 15

Special uses of Concrete Pipes

15.1 SUBMARINE OUTFALLS OF CONCRETE PIPES

15.1.1 Historical

Submarine pipeline involve more complexity and challenge to the engineer and contractor than any other type of pipeline construction. Submarine pipeline construction is not a new development, but the techniques, equipment, and the sophistication of pipe and joint design have evolved through the years. One of the earliest examples is a cooling water intake constructed for the public service Electric and Gas Company at Jersey, in 1910. This 1370 mm diameter reinforced concrete pressure pipeline, constructed with flanged joints, is still in service.

Today there are thousands of submarine installations and more are being constructed at an accelerated rate. These lines, located in fresh, brackish and salt water, cover a wide range of application including-

(a) Water or sewer line crossing under rivers, theriatrics or inland lakes;

(b) Industrial cooling water intake and discharge lines;

(c) Treated sewage outfalls;

(d) Intake line for industrial or municipal water supply.

Material for the Conduit

Potential loads and the surroundings to which the pipe line will be subjected are

(i) Internal Pressure – Hydrostatic pressure of 1 to 2 kg/cm^2.

(ii) External Load (water and over burden) – If the pipeline is floating then the question of buoyancy does not arise, otherwise the external load to take care of the up-lift pressure due to submergence, when the pipe is empty need to be considered.

(iii) Corrosive surrounding – Pipe is constantly subjected to seawater, where high chloride environment highly corrosive to ferrous metals is expected.

A chloride concentration of 20,000 parts per million is considered to be acting on the steel.

Considering all the above factors the choice fall on the following materials.

(A) Concrete Pipe

(i) Without steel Cylinder – For diameters 100 mm to 5400 mm. pipes to be laid underground with a cover of filling over it. The pipes are jointed by flexible rubber ring joint.

(ii) With steel Cylinder – For diameter 100 mm to 540 mm. Pipes to be laid underground with a cover of filling over it. The pipes are jointed by flexible rubber ring joint, for higher internal pressure than above.

(B) Steel Pipe

With protective lining and out-coating to cover sizes from 300 mm diameter to 1200 mm diameter. The pipe is generally laid on the bed of the sea. The pipes are jointed by welding on the barge over which protective coating is applied and then the whole pipeline is pulled on the sea bed. Few field joints can be done by welding under water if required. The diameter range is limited as the pipe has to take a bend during pulling.

(C) P.V.C. Pipe

Maximum diameter up to 450 mm is common. The pipe is usually floating and it is anchored at various places to the buoys. Sometimes it is also laid on the bed of the sea by providing anchor blocks.

In this paper, an attempt is made to present information about large diameter outfalls made of concrete pipes, which are usually without steel cylinder. Hence, the relative merits of various pipe materials are not discussed. The above remarks are only broad indications.

Concrete Pipes for ocean outfalls

The most challenging and complex sub aqueous lines are the major ocean outfalls. Some of these, such as 'Hyperion Outfall' in California in 1910 has been monumental engineering achievements. For such projects the sophistication employed in pre contracts investigations have been consistent with the magnitude of the projects themselves, and involve a high degree of oceanographic technology. Environmental considerations relating to proper dispersion of the effluent, protection of beaches and the marine ecology require a through analysis of the total characteristics of each project.

The two major sections of construction of an ocean outfall are–

1. The on-or-near-shore section and
2. The off-shore section.

Each of these sections has its own specific problems and construction techniques. Initially the pipeline has a bigger diameter, which after the safe diffusers fitted on it, and subsequently the diameter goes on reducing.

On-or-Near – Shore sections

These sections are affected by both wave and tidal action and are the most critical from the point of potential external forces and influences on the pipe. For this reason tense sections are always buried. Conventional construction procedures are followed on the dry land section until the beach and surface area are reached. From this point a temporary trestle or pier is usually built to handle the excavating and pipe handling equipment.

Off-shore sections

These are no clear cut answers to the best method for constructing offshore pipelines. For large enough projects, special equipment can be justified, such as "special built ferrys" a giagantic 4-legged, 5500—ton mobile laying platform capable of handling up to 60 meters set of pre-assembled large diameter pipes. The rig consists of four legs or towers which can hydraulically jack the platform out of water, so that it is supported by the sea floor for positive control in laying the pipe assembly. Retracting the legs allows the platform to be floated to the next position, where it is again jacked up.

Construction frames or towers are also used from 8 meters to approximately 24 meters. These rest on the sea floor to provide a stable laying platform and are moved with a derrick barge. For depths beyond 24 meters, present methods often utilize floating equipment, Long assemblies of pipes can be handled this way by use of pontoons or other means of closely controlling the lowering of the pipe. Control techniques for line and grade beyond feasible limits for shore control or bottom supported laying platform frequently consists of survey towers, specially anchored buyos and guyed risers on the pipe.

Manufacture of concrete pipes

Pipes up to 3000 mm diameters are made till this day spinning. Large diameters are cast vertically between smooth steel forms set in a special base ring to hold forms and reinforcement, concentric at the base. An accurate mechanical slotted working platform maintains concentricity of the forms and reinforcement at the top. It also provides a surface from which concrete is distributed uniformly between the forms. As high grade concrete is poured slowly, the forms are vibrated to produce densely compacted concrete with a smooth finish.

Reinforcing cages

Cages are fabricated from coils of hog rolled steel rods wound helically on a collapsible mandrel. The cages are supported longitudinally by steel rods, around the cage. One or more cages are used depending upon the diameter of the pipe and the conditions of the design.

Joints

The spigot is designed to accommodate two 'o' rubber rings which serve as pressure seals between the jointing surfaces. This also serves as a method of testing joints closures underwater. A 12 mm diameter round model tube is encased in the spigot with one opening between the rubber rings and other on the outside surface of the pipe near the spigot end. Air is directed into the space between the rings and a pressure gauge indicates any leakage.

Excavation

The excavation is usually done by a chanshell bucket and the excavated stuff is carried away by barges. The bedding and backfilling material is usually brought to the site from the shore by barges. The precise placing of bedding and backfilling is done via a pipe with at the top.

Bedding

It depend upon the bearing capacity of the sea bed, sometimes rock is met with in such cases very little bedding is required. Generally bedding of aggregate or gravel of thickness varying from 300 to 1000 mm is adequate. In exceptional cases concrete is also required.

Backfilling

This depends upon the relative level of sea bed and the pipeline. But, backfill of rock graded with coarse aggregate for depths varying from 1.5 to 2.5 meters are not uncommon.

Jointing system

In recent years the hydro-pull systems has been used to join pipes under water. A bulched with a high volume pump built into it is attached to the end of the pipe to be laid. The pump evacuates water at a rate sufficient to create a pressure differential which is grate enough to force the joint together.

The lengths of the pipe which may up above water depend largely on the capacity of handling equipment. Instances of assemblies as great as 56 metes consisting 8 pipes of 7 meters each are not uncommon. Such intermediate joints made above water may be conventionally jointed, depending on the required flexibility of the line and the method of handling the assembly. These assemblies are then lowered as a unit using specially designed and fabricated frames and the underwater joints made.

Divers are used to guide and monitor the placement of pipe sections with 3 to 4 hours time needed per joint.

Joint testing in field

After the pipe assembly is installed and jointed to the previously platform and connecting it to inlet of the monel tube. During installation of pipes it is ensured that the inlet to the monel tube is always up, to facilitated connection of air. Air is pumped from the compressor and pressure gauge is watched. At this stage pressure is applied between the two rubber rings. If the jointing is correct, pressure gauge will remain study. During this operation the help of divers is most essential. If the joint is satisfactory, filling is undertaken and further work proceeds.

Buoyancy

Potential floatation of a submarine pipeline must always be considered if the pipeline can be emptied. Heavy walled pipe, concrete weights, rubble, shingle or combination of these can be employed to overcome the buoyancy of an empty pipeline. Dead-weighting is generally the most practical method of anchoring off-shore pipelines, coarse backfill material being most commonly employed.

Quality control

Quality control at every stage of production and installation is to ensure that the pipe wall is durable, impervious and the joint is water tight. These characteristics are particularly important in subsequent pipes, since its installation is amongst the most complex, costly and limited to few seasons. Consequent and further repairs, replacement or maintenance of the line will be proportionately difficult and expensive.

The most important being the embedment of steel in dense impervious wall of concrete, whose smooth surfaces will neither corrode nor tuberculation throughout a useful life conservatively estimated at more than 100 years.

It should mainly aim at checking cover over steel; the dimensions of socket, spigot and their ovality.

15.1.2 Case Studies

(1) San Onofore's offshore Pipe (1976)

Purpose - For the disposal of cooling water used by generating units.

Total Length of the Pipeline- 8200 feet

Pipe Details –

Diameter	- Varying from 18 feet down to 10 feet.
Wall Thickness	- 1 foot
Length of the pipe	- 24 feet
Weight of each pipe	- 110 tons
Total No of pipes used	- 880 they were caste and transported to the site in specially
Diffusers	- 126 designed equipment.
Water Depth	- 50 feet

In the portion of the sea near the coast, the pipes were laid by construction trestles and beyond that the pipes were laid by jack up Barge.

Jack-up barge

With an open center hull i.e., 18 feet deep, 90 feet wide long with a jack leg system. It weighted 3000 tons and was supported by 4 tons. 8 feet diameter legs, each leg had a jacking capacity of 1000 tons.

Installation of pipe in sea

The water was no where more than 50 feet deep. So the entire project could have built from a trestle, but to avoid erecting a barrier to coastal ship and boat traffic, specifications called for floating equipment.

The intake and discharge line was side by side and were laid from the trestle as far as possible. Beyond that the jack-up barge took over.

The pipe sections made in yard set up across inter state from the project and were brought to the construction site by latest version of pipe mobile. The pipe is laid between the rails at the on shore end of the trestle where it could be picked by jack up barge.

The pipes are then carried through the jack-up barge to the point of placement. The jack-up barge is adjusted exactly on the required alignment of the pipeline and jacked up. Pipe is then lowered through the central opening in the barge. Before the pipe entered water necessary ropes were tied to the pipe to enable it to pull it for fixing it to the pipe already laid. When the pipe is at proper elevation, the wire fops are tightened to make the joint. Manipulating the tension in the wire rope orients the pipe properly.

Diffusers

Diffuser port blocks are located at 40 feet centers. The diffuser heads are elevated approx. 8 feet above sea floor. The individual diffuser block weighed 68, 63 and 60 tons depending on diameter of pipe on which they were installed.

Positioning of diffuser heads along each discharge line minimizes the effect on sea temperature of the slightly warmer returned to the sea.

(2) Orange country Outfall (1969)

Purpose - Sewage Outfall

Total Length of the Pipeline - 2700 feet

Pipe Details –

Diameter   - Varying from 10 feet to 6 feet.

Weight of each pipe   - 67 tons.

Water Depth   - 198 feet.

Laying platform

After removing the engine of a 440 foot long ship, the hull served as a platform. The ship was equipped with 175 tons capacity crane, 2 drum winches, twin deck mounted 20 cubic yard aggregate hopper, decompression chamber for divers and other communication facilities.

Trenching, bedding and backfilling

This is achieved by 6 cubic yard capacity clamshell bucket aboard a 2000 tons barge having length equal to 200 feet.

Installation of pipe in sea

Maximum depth of water where pipeline was to be laid was 198 feet. The work of laying in this portion was simplified by using massive steel framework. The frame called 'Horse' was quipped with 4-controllable lengths legs so that the pipe, slug under the frame can be set on grade. The pipe was pulled into previously laid section with cables leading to the ship mounted engine driven winches with operators following instructions telephoned to the laying platform by divers. Slings were released after joining and the 'Horse' raised and lowered by crane was brought back aboard, set on deck and supplied with next pipe section. The cycle is completed rapidly. When operations are not hampered by poor underwater visibility, surface weather or equipment servicing requirements. The crew included about 20 men during offshore operations including 3 divers.

Bedding

1-3 feet deep gravel bedding. Aggregate backfill up to 5 feet height was specified over the pipe in sea and large rocks were placed over pipe close to shore up to the pipe section.

Concrete pipe

Concrete pipe has a socket and spigot joint. The spigot was designed to accommodate two neurons 'o' rings which serve as a pressure seal between the joints and which provides a method for testing joint underwater. A 0.5" round 'Monel Tube' is enclosed in the spigot with one opening between the rings and second outside surface of the pipe near spigot end. Air is directed into space between rings and pressure gauge reading indicates any leakage.

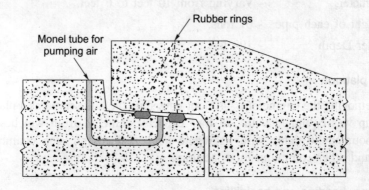

Fig. 15.1 Joint for ocean outfall (Two rubber rings)

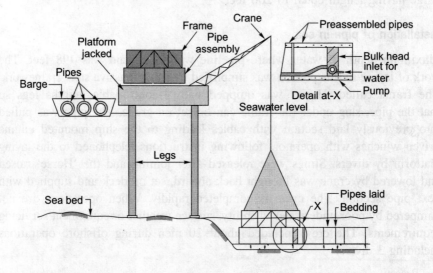

Fig. 15.2 Sketch showing jointing system for pipes under sea

Rubber ring joint was introduction in 1934 in place of caulked joint. This was a, major improvement in many respects as it greatly enhanced the advantages of reinforced concrete pipe for widening range of pressure. Initially

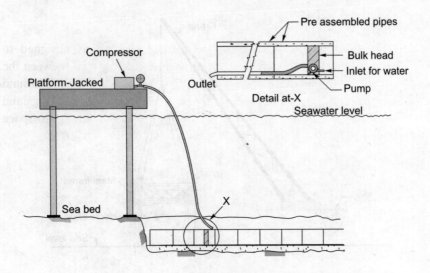

Fig. 15.3 Arrangment showing testng of joint at field

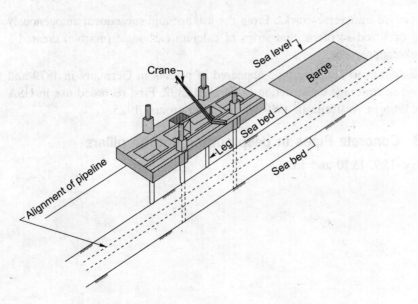

Fig. 15.4 The jack-up barge for san onofre's off shore outfall

the Pressure pipes were jointed with steel collars. Those were mortar caulked, to joint about every 4th joint. This was marginally satisfactory for buried pipes but in portion exposed to severe frost, hot sun and drying winds, even if all the joints were of lead, they would allow only small faction of expansion and contraction as afforded by rubber ring joints. The resulting longitudinal

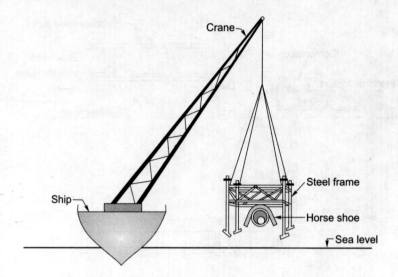

Fig. 15.5   Installation of pipe for orange sea outfall

forces caused transverse cracks. From the leakage and subsequent autogenously healing occurred so often, that strips of calcium carbonate (marble) created a zebra-like appearance.

The reinforced concrete pipe appeared in patents in Germany in 1879 and was used in sewerage construction in Paris in 1892. First recorded use in USA was in 1906 in Australia in 1910 and in India around 1925.

15.1.3   Concrete Pipes in deep cutting and on pillars

See Figs. 15.9, 15.10 and 15.11

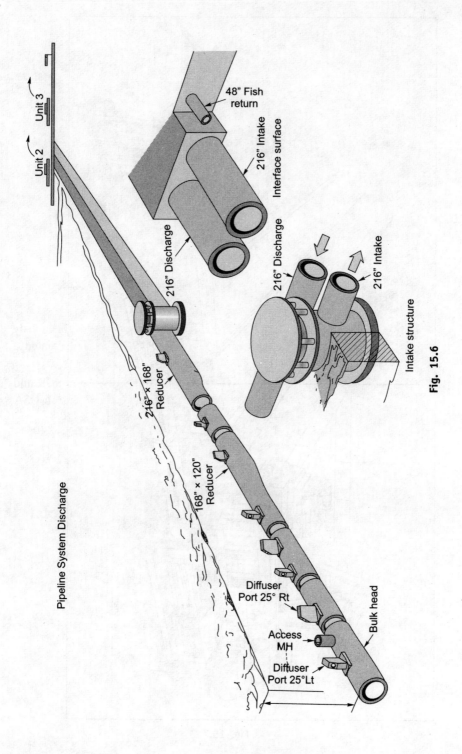

Fig. 15.6

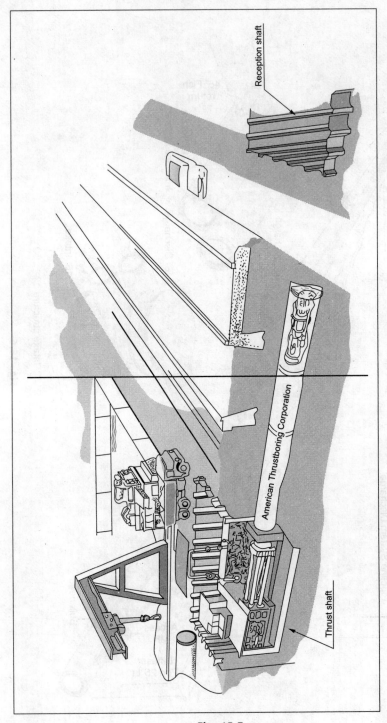

Fig. 15.7

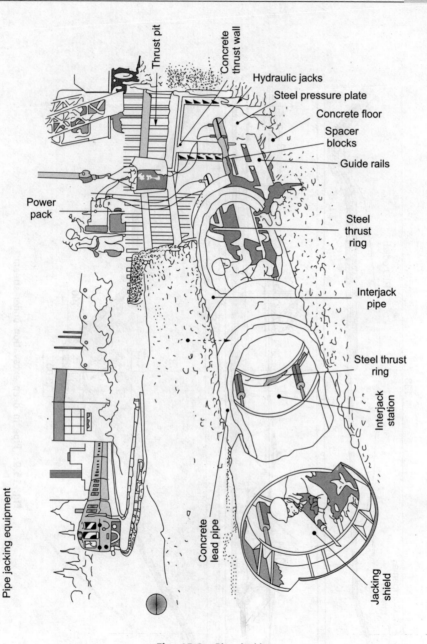

Fig. 15.8 Pipe jacking

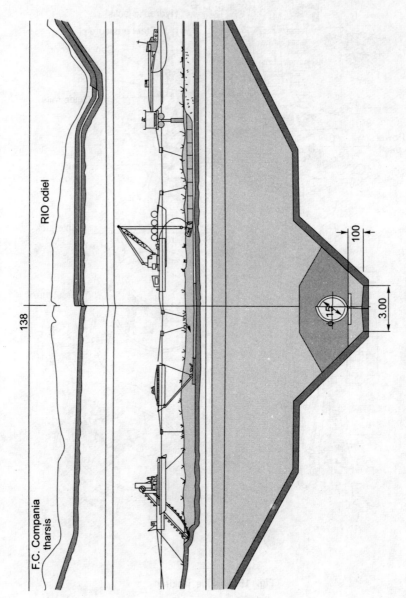

Fig. 15.9   Pipe in deep excavation under stream

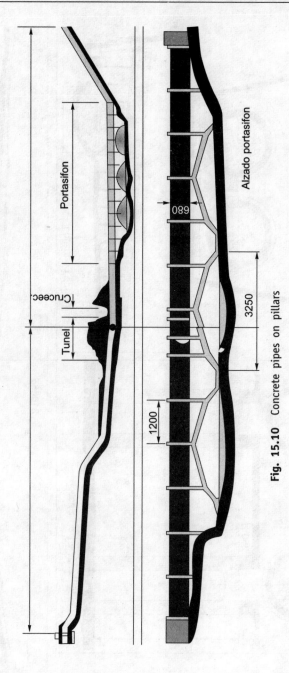

Fig. 15.10 Concrete pipes on pillars

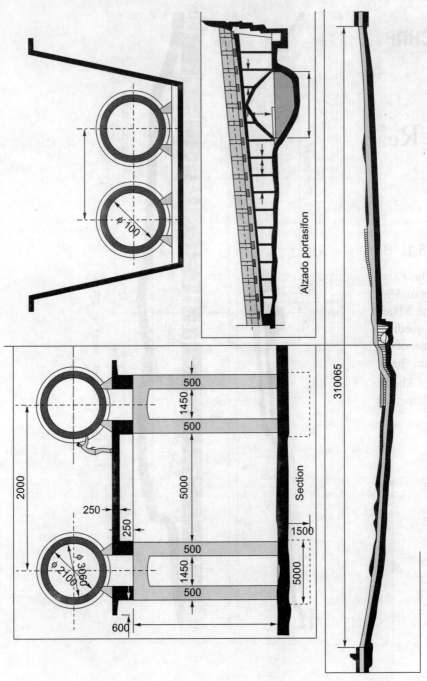

Fig. 15.11 Use of concrete pipes on pillars

CHAPTER 16

Results of Testing in 1997 of 75 Years Old Concrete Pipe

16.1 BACKGROUND

The City of Winnipeg, Manitoba, Canada, receives all of its drinking water from Shoal Lake, which is approximately 140 km (87 mi) east of the city at the Manitoba/Ontario border, by a pipeline system known as the Shoal Lake Aqueduct. Constructed between 1917 and 1919, the Branch I Aqueduct provides one of two final links between the Deacon Reservoir on the city's outskirts and the urban distribution system.

The Branch I Aqueduct consists of 15,210 m (49,900 ft) of 1675 mm (66 in.) diameter precast reinforced concrete pipes from the Deacon Reservoir to the

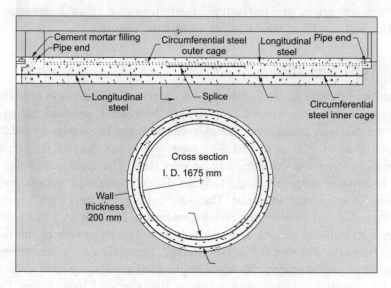

Fig. 16.1 Test pipe sections

east bank of the red River, 314 m (1030 ft) of 1270 mm (50 in.) diameter concrete –encased cast iron pipes under the river, and 3475 m (11,400 ft) of

1220 mm (48 in.) diameter precast reinforced concrete pipes from the west bank of the Red River to the McPhillips Reservoir. The precast concrete pipes were made in 3.05 m (10 ft) lengths for the 1220 mm diameter pipe and in 2.44 m (8 ft) lengths for the 1675 mm diameter pipe.

The Branch I Aqueduct was originally designed to operate at a full capacity of 250 ML/day (66 MGD), with maximum working heads of 12.2 and 229 m (40 and 95 ft) for the 1675 and 1220 mm diameter pipes, respectively. However, since its commission in 1919, the pipeline has been operating at a reduced capacity of 115 ML/day (30.4 MGD). Recent water demand studies for the City of Winnipeg have identified the need for the Branch I Aqueduct to operate at its maximum design capacity. Corners exist, however, as to the ability of the pipe to operate at increased pressures after 75 years of continuous wear and tear on the pipeline in the sulfate-rich soil.

Approximately 460 m (1509 ft) of the original 1675 mm diameter pipe were replaced in 1966 to accommodate the construction of the Red River Floodway. Among the nearly 200 pipe sections removed, four were recently located. As part of an overall structural assessment, full-scale hydrostatic pressure testing was conducted on the salvaged pipe sections to confirm the original design strength of the pipe, to determine the impact of reduced wall thickness by simulating external sulfate attack on the concrete, and to provide experimental data for verification of a finite element analysis model.

Presented herein are test setups, instrumentation, data acquisition, and experimental results of the full-scale hydrostatic testing.

16.2 TEST SET UP

Of the four salvaged sections of the 2.44 m (8 ft) long 1675 mm (66 in.) diameter pipe, one section was selected for full-scale hydrostatic pressure testing. The 1675 mm diameter precast concrete pipe was designed to withstand a maximum working head of 12.2 m (40 ft). The pipe wall is 200 mm (8 in.) thick and contains an inner and outer reinforcing steel cage. The inner reinforcing steel cage consists of three 6.4 mm (0.25 in.) diameter stranded smooth wires at 100 mm (1 in.) center to center. The outer reinforcing steel cage consists of 12.7 mm (0.5 in.) twisted square bars spaced at 114 mm (4.5 in.) center to center with reduced spacing near the bell end. The lap length of the twisted square bars was 380 mm (15 in.) on average. Lap spices are circumferentially staggered at approximately 50 deg. Longitudinal reinforcing steel consists of ten 12.7 mm (0.5 in.) plain square bars in the outer and inner layers. Details on the reinforcing steel are shown in Fig. 16.1.

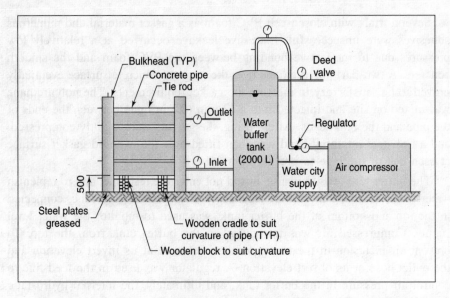

Fig. 16.2 Design of test set up

Prior to testing, no cracks or other defects were observed on either the exterior or interior pipe surface, except for minor damage to the pipe ends due to rough handling and transportation over the years. A diamond-wire saw was used to trim the ends of the pipe square and smooth. Prior to cutting, one end of the pipe was x-rayed to locate the reinforcing steel and to identify a cut line that would avoid both the inner and outer circumferential reinforcing steel. The remaining saw cut was located by exploratory drilling from the end of the pipe. The length of the trimmed test pipe was 2088 mm (82 in.), or about 350 mm (14 in.) shorter than its original length.

The test set up consisted of the concrete test pipe, concrete bulkheads for both ends of the pipe, a 2000 L (528 gal.) water buffer tank, and a 1400 kPa (230 psi) air compressor (Fig. 16.2). The pipe was supported by a timber cradle shaped to match the curvature of the pipe. The concrete bulkheads (2650 × 2650 × 450 mm (104 × 104 × 18 in.)) were clamped to the pipe by eight 38 mm (1.5 in.) diameter threaded tie rods. Two sets of greased steel plates were used beneath each bulkhead to facilitate positioning of the test set up. Access holes were provided through one bulkhead for inlet and outlet ports instrumentation leads.

The design and construction of the gasket between the ends of the pipe and the concrete polyurethane proved to be the most challenging aspect of the test set up. The gasket was required to provide a flexible watertight seal that would minimize the effects of longitudinal compression of the pipe from clamping the bulkheads, and at the same time, withstand maximum test pressures.

Several trials with closed-cell PVC foam as a gasket material and numerous adhesives were unsuccessful. Excessive leakage occurred at a relatively low pressures due to ineffective bonding between the PVC foam and the smooth concrete. A two-part polyurethane over the roughed concrete surface eventually proved to be satisfactory in meeting the gasket requirements. The polyurethane was mixed on site and injected into a 17 mm (0.7 in.) gap between the ends of the pipe and the bulkheads. After curing, the gaskets were slightly compressed, and a flat steel retaining band was then fitted over the exposed gasket surface at each end of the pipe to prevent blowout.

The buffer tank was to insure air did not enter the test pipe, and to replenish water to the test pipe in case of leakage. A clear plastic sight tube, connected to the top and bottom of the buffer tank, was used to monitor the water level supply. Compressed air was introduced to the buffer tank from the top. To prevent air inclusion in the test pipe, the inlet hose at its invert elevation and the outlet hose at its obvert elevation. A regulator was used in the feed line to control air pressure in the buffer tank, and ultimately, the internal hydrostatic pressure. (Fig. 16.3 shows the actual test set up minus the air compressor.)

Two vibrating wire displacement sensors were installed inside the test pipe to measure changes in the diameter in the horizontal and vertical planes. The sensors were located at the center of the pipe where the influence from clamping forces would be negligible. A vibrating wire piezometer was installed at the middle inside surface of a bulkhead to monitor internal pressure and temperature to apply thermal corrections to observed displacements.

Cables from the three devices were passed through ports in the bulkhead and then connected to a digital data logger via a switch box. Frequency readings of the vibrating wire sensors were stored in the data logger and were later downloaded to a computer that processed the data according to the sensors'

Fig. 16.3 Actual test set up

Fig. 16.4 Test pipe with 38 mm of concrete removed and reinforcing steel exposed

individual calibration curves. Dial gages were also mounded on the pipe to confirm diametric displacements of the pipe, and to measure the movements of the bulkheads.

Although there are no standard procedures for this particular type of hydrostatic testing, ASTM C 361M1993, which outlines a standard low pressure hydrostatic test for joint leakage of precast concrete pipes, was followed whenever possible. The entire test was divided into three phases.

Phase 1 Testing the pipe to 220 kPa (32 psi) pressure, which represents 1.2 times the maximum anticipated pressure, which includes a 50% allowance for transient pressure of the pipeline under the full flow capacity. Three cycles of loading and unloading were carried out.

Phase 2 Testing the pipe to a maximum pressure of 345 kPa (50 psi), which corresponds to about twice the maximum anticipated pressure (inclusive of transient pressure). Two cycles of loading and unloading were carried out.

Phase 3 Layers of external concrete were removed by water jetting to simulate sulfate attack. Concrete was removed from the top two-thirds of the pipe, which was considered representative of a worst-case scenario for sulfate attack. Testing was carried out to a pressure of 220 kPa (32 psi) for 10, 20 and 38 mm (0.4, 0.8 and 1.5 in.) of concrete removal. At a depth of 38 mm (1.5 in.) into the concrete pipe, the outside surface of the outer reinforcing steel was exposed. Two loading cycles were carried out for each stage of concrete removal.

Internal pressure was then increased to 345 kPa (50 psi), and ultimately, to 600 kPa (87 psi) — equivalent to a factor of safety of 3.3. Limitations of the test apparatus and instrumentation prevented increasing test pressures beyond 600 kPa (87 psi). Test pressures and degrees of Test results a are summarized in Table 16.1.

Prior to the first test, the pipe was presoaked for 48hr at 35 kPa (5 psi) pressure. During testing, water pressures were typically increased from 0 kPa to the maximum test pressure in increments of 34 kPa (5 psi), with each increment held for 10 min. Readings from the vibrating wire devices and external dial gages were recorded at the beginning and end of each 10 min holding period. The maximum test pressure was held for 20 min and the pressure was incrementally decreased in the same fashion. Each cycle of testing was separated by at least 15hr to allow the pipe to recover dimensionally. The exterior of the test pipe was visually examined for evidence of cracking during all tests.

16.3 RESULTS AND DISCUSSION

Horizontal and vertical diameter changes were averaged to account for slightly imperfect roundness of the test pipe, and are presented in Fig. 16.5 through

Table 16.1 Summary of test phases and pressures

Test phase	Test identifier	Test pressure		Factor of safety	Thickness of concrete removed, mm
		kPa	psi		
1	32-1	220	32	1.2	0
	32-2	220	32	1.2	0
	32-3	220	32	1.2	0
2	50-1	345	50	1.9	0
	50-2	345	50	1.9	0
	50-3	345	50	1.9	0
	32-4	220	32	1.2	10
	32-5	220	32	1.2	10
	32-6	220	32	1.2	10
	32-7	220	32	1.2	20
	32-8	220	32	1.2	20
	32-9	220	32	1.2	20
3	32-10	220	32	1.2	38
	32-11	220	32	1.2	38
	32-12	220	32	1.2	38
	50-4	345	50	1.9	38
	50-5	345	50	1.9	38
	87-1	600	87	3.3	38

16.8. Data from all test replicates during loading and unloading are included in each figure. Also included are the theoretical lines calculated based on uncracked sections. In the theoretical calculation, the area of reinforcing steel was transformed into an effective concrete area in the circumferential direction. Strength of the reinforcing steel and the concrete was determined from samples taken from the test pipe after completion of the testing.

Figure 16.5 shows the results of Phase 1 testing. No concrete was removed from the pipe in this test phase. As shown in the figure, the pipe behaved linearly during the test, and the test data are close to the theoretical line. The maximum diameter change corresponding to 220 kPa (32 psi) internal pressure was approximately 0.051 mm (0.002 in.). The slope of the regression line was 4220.56, 4.4% less than its theoretical slope.

The results of Phase 2 testing at 345 kPa (50 psi) are shown in Fig. 16.6. Although a linear relationship exists between internal pressure and diameter

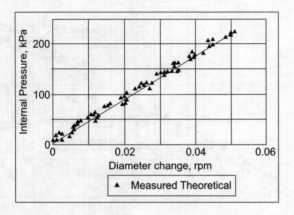

Fig. 16.5 Result of Phase 1 testing; test pressure = 220 kpa

·change, the pipe started to respond bilinearly when the pressure reached 300 kPa (43.5 psi). When unloading, no permanent deformation (creep) was observed even after the non-linear response of the pipe. The maximum diameter change corresponding to 345 kPa was approximately 0.09 mm (3.5 mils). The slope of the regression line was 4080.56, 7.6% less than its theoretical slope.

The results of Phase 3 testing with 10, 20 and 38 mm (0.4, 0.8 and 1.5 in.) concrete layers removed from three-quarters of the exterior surface are presented in Fig. 16.7. The large scattering of the data for the 10 mm concrete removal was

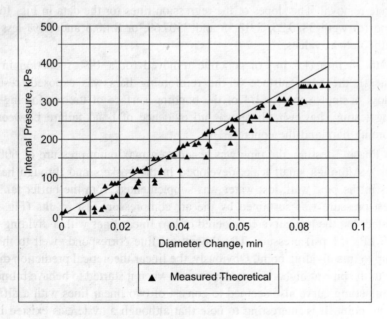

Fig. 16.6 Result of Phase 2 testing; test pressure = 345 kpa

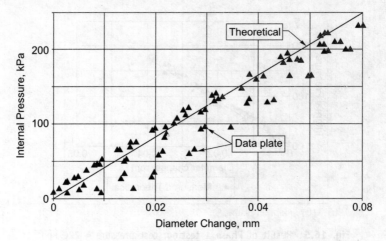

Fig. 16.7(a) Test Pressure = 220 kpa, Concrete Removal = 10 mm

possibly due to the non-uniform concrete removal in the initial stage. A linear regression of the test data gave an R-square of 0.94, 0.98 and 0.98 for the 10, 20 and 38 mm of concrete removal, respectively. Statistically, the relationship between the pipes diameter change and its internal pressure was still linear for all three cases for a test pressure of 220 kPa (32 psi). The maximum diameter change corresponding to the test pressure of 220 kpa was 0.06 mm (2.4 mils) for the 10 and 20 mm concrete removal, and 0.07 mm (2.8 mils) for the 38 mm concrete removal. The slopes of the regression lines for the data in Fig. 16.7(a), (b) and (c) were 3820.25, 3918.55 and 3507.69, or 8.1,0.5 and 0.1% less than their theoretical values.

With 38 mm (1.5 in.) of concrete removed (or a 19% reduction in wall thickness), the outer surface of the reinforcing bars was exposed, resulting in a loss of approximately 1/4 of the bonding surface of the reinforcing bars. During testing, however, there was no evidence of bond failure between the reinforcing bars and the concrete.

In Phase 3 testing, the pipe was tested to a maximum pressure of 600 kPa (87 psi). Although small leaks developed in the gaskets and through hairline cracks in the pipe wall, lost water was supplemented from the buffer tank and the test pressure was sustained by the air compressor. Test results (Fig. 16.8) indicated that the test curve considered of two linear lines with a dividing point at 350 kPa (51 psi) pressure. The theoretical line corresponds well to the test data up to this dividing point. Obviously, the linear theoretical prediction did not apply to the pipe at higher internal pressures when it started to behave bilinearly. The unloading curve also seemed to consist of two linear lines with a different dividing point. It is interesting to note that although a hysteresis existed in the loading-unloading cycle, no permanent deformation (creep) resulted after the

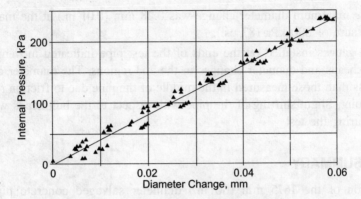

Fig. 16.7(b)   Test Pressure = 220kpa, Concrete Removal = 20 mm

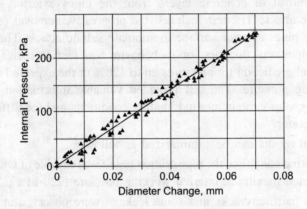

Fig. 16.7(c)   Test Pressure = 220 kPa, Concrete Removal = 38 mm

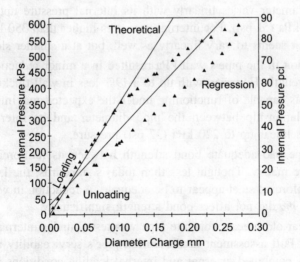

Fig. 16.8   Result of Phase 3 testing with 38 mm of concrete removed; test pressure = 600 kPa

test. The maximum diameter change was 0.28 mm (0.01 in.) at the maximum test pressure of 600 kPa (87 psi).

Dial gauges installed on the ends of the test pipe indicated movements of the bulkheads and diameter changes in the end regions. The diameter changes were less than those measured in the middle of the pipe due to friction restraint of the joint. Slight shifting of the pipe with respect to the bulkheads was also noted during the test.

16.4 SUMMARY

A section of the 1675 mm (66 in.) diameter salvaged concrete pipe was hydrostatically tested to various internal pressures. The test included the progressive removal of concrete layers from the pipes exterior surface to simulate sulfate attack. The test included the progressive removal of concrete layers from the pipes exterior surface to simulate sulfate attack. The effect of reduction the pipes wall thickness on its behavior was seen to correspond well to the theoretical prediction for pressures up to 120% of the expected maximum internal working pressure. The test provided valuable information regarding the pipes response to various internal loading conditions and for different wall thickness reductions.

The test and results can be summarized as follows:

1. In its existing condition, the pipe section tested was capable of withstanding the expected maximum internal working pressure (220 kPa psi)

2. Although hairline cracks and some leakage were observed at pressures exceeding 345 kPa (50 psi), no structural failure occurred.

3. Pipe diameter varies linearly with its internal pressure upto a pressure of 350 kPa (51 psi). For internal pressures higher than 350 kPa, the pipe diameter seems to vary linearly as well, but at a smaller slope.

4. Reduction in the pipe's wall kPa resulted in a minor reduction in overall pipe strength. However, with up to a 19% loss in wall thickness, the test pipe was capable of functioning under the expected maximum pressure. The relationship between the pipe diameter and its internal pressure remains linear up to 220 kPa (32 psi) pressure.

5. The pipe had adequate bond strength between its reinforcing bars and concrete matrix. Thought less than today's standards, the lap lengths of the reinforcing steel appear to be adequate. A reduction in wall thickness up to 19% did not affect bond strength significantly.

The 75-year-old precast concrete pipe was tested under internal hydrostatic loading only. Full assessment of the buried pipe's serviceability and integrity under various combined external and internal loading conditions is still under way.

Appendix

JSW CEMENT LTD.
TEST CERTIFICATE
GRAUND GRANULATED BLAST FURNACE SLAG

IS NO	Characteristics	Requirement as Per BS: 6699	Test Result
1	Fineness (M2/Kg)	275 (Min)	407
2	Insoluble Residue (%)	1.5 (Max)	0.57
3	Magnesia Content (%)	14 (Max)	9.00
4	Sulphide Sulphar (%)	2 (Max)	0.47
5	Sulphide Content (%)	2.5 (Max)	0.10
6	Loss on Ignition (%)	3 (Max)	0.20
7	Manganese Content (%)	2 (Max)	0.05
8	Chloride Content (%)	0.1 (Max)	0.01
9	Glass Content (%)	67 (Min)	88
10	Moisture Content (%)	1 (Max)	0.12
11	Chemical Moduli		
A	$CaO + MgO + Sio2$	66.66 (Min)	74.9
B	$CaO + MgO/Sio2$	>1.0	1.32
C	CaO/SiO_2	<1.40	1.04

Note: There is Indian Standard on GGBS. For reference BS specification is given. Granulated Slag used for Steel Ltd., GBS conform to IS 12089 : 1987

Week no :37
07-0.9-2009-13-09-2009

Authorised Signatory

Index